T0182072

Raw Materials for Future Energy Supply

Friedrich-W. Wellmer · Peter Buchholz
Jens Gutzmer · Christian Hagelüken
Peter Herzig · Ralf Littke
Rudolf K. Thauer

Raw Materials for Future Energy Supply

With Contribution from Gerhard Angerer, Berit Erlach,
Tobias Kracke
With Support of Marie-Christin Höhne

 Springer

Friedrich-W. Wellmer
Academy of Geosciences and
  Geotechnology
Hannover, Germany

Peter Herzig
GEOMAR Helmholtz Centre for Ocean
  Research
Kiel, Germany

Peter Buchholz
German Mineral Resources Agency (DERA)
Federal Institute for Geosciences and Natural
  Resources (BGR)
Berlin, Germany

Ralf Littke
Faculty of Georesources and Materials
  Engineering
RWTH Aachen University
Aachen, Germany

Jens Gutzmer
Helmholtz-Zentrum Dresden-Rossendorf
  (HZDR)
Helmholtz Institute for Resource
  Technology (HIF)
Freiberg, Germany

Rudolf K. Thauer
Max Planck Institute for Terrestrial
  Microbiology
Marburg, Germany

Christian Hagelüken
Umicore AG & Co KG
Hanau, Germany

Translated by Duncan Large

Analysis published in the series "Energiesysteme der Zukunft" (Energy systems of the future, ESYS) of the German scientific academies: acatech—National Academy of Science and Engineering, German National Academy of Sciences Leopoldina, Union of the German Academies of Sciences and Humanities, Munich, Berlin 2016 (Status November 2015).

ISBN 978-3-030-08203-1   ISBN 978-3-319-91229-5   (eBook)
https://doi.org/10.1007/978-3-319-91229-5

Preface

The energy transition will fundamentally change our consumption of natural resources. As solar and wind energy increasingly meet the energy requirements in Germany, the demand for coal, oil, and gas will decrease over the long term. At the same time, the consumption of metallic raw materials necessary for the manufacture of energy-efficient wind and solar generating facilities, batteries, and hydrogen storage, and other systems related to the energy transition, will increase. In addition to copper, cobalt, and nickel, the rare-earth elements and other high-technology metals will be required and must, to a large extent, be imported. German industry has become dependent on these new raw materials, and in 2009, the associated risks were accentuated by the drastic increases in the prices of the rare-earth elements.

Is the implementation of the energy transition threatened by a lack of natural resources? The authors of this study conclude that sufficient supplies of metals, fossil fuels, and bioenergy can be safeguarded. Promising technological approaches are described—from expansion and improvement of the recycling capacity to the exploitation of completely new sources of raw materials from the deep ocean.

The interrelationships with the global commodity markets are also evaluated. What are the critical raw materials? How quickly can the market react to a sudden increase in demand? How will the global demand develop during the coming decades? Answers to these questions assist both in the timely identification of warning signs indicative of risks in the supply chain as well as in the development of alternative strategies.

Finally, this study also examines the ecological and social impacts of mining. If the transition to "green" energy should indeed result in greater sustainability, then, to be consistent and consequent, the exploitation of the necessary natural resources must also be taken into consideration. A reliable, affordable, and environmentally and socially acceptable supply of natural resources is therefore a key component for a successful energy transition.

The authors like to thank Gerhard Angerer, Tobias Kracke, and Berit Erlach for their valuable contributions and GEOMAR Helmholtz Centre for Ocean Research Kiel as well as GFZ German Research Centre for Geosciences Potsdam for financial support.

Kiel, Germany   Prof. Dr. Peter Herzig
                                              Chairman of ESYS Working Group "Resources"

Contents

Abbreviations and Units

AFMAG   Audio frequency magnetics
ATP   Adenosine triphosphate
BEV   Battery electric vehicle
BDI   Federation of German Industry (Bundesverband der Deutschen
             Industrie e.V.)
BGR   Federal Institute for Geosciences and Natural Resources (German
             Geological Survey, Bundesanstalt für Geowissenschaften und
             Rohstoffe)
BGS   British Geological Survey
BMBF   Federal Ministry of Education and Research (Bundesministerium
             für Bildung und Forschung)
BMEL   Federal Ministry of Nutrition and Agriculture (Bundesministerium
             für Ernährung und Landwirtschaft)
BMELV   Federal Ministry of Nutrition, Agriculture and Consumer Protection
             (Bundesministerium für Ernährung, Landwirtschaft und
             Verbraucherschutz)
BMW   Bayrische Motorenwerke AG (German automobile manufacturer)
BMWi   Federal Ministry for Economics and Energy (Bundesministerium
             fürWirtschaft und Energie)
BMUB   Federal Ministry for Environment, Nature Conservation, Building
             and Nuclear Safety (Bundesministerium für Umwelt, Naturschutz,
             Bau und Reaktorsicherheit)
BÖR   Council for Bio-economics (BioÖkonomieRat)
BRIC States The BRIC States include the rapidly developing industrial countries
             Brazil, Russia, India and China
CAS   Chemical Abstracts Service
CCS   Carbon Capture and Storage, technical term for technologies
             focused on the separation and storage of carbon dioxide
CONNEX   G7 initiative to support developing and emerging countries in
             complex contract negotiations

CPI	Corruption Perceptions Index
CRB	Commodity Research Bureau
CSR	Corporate social responsibility
CTC	Certified Trading Chains
DERA	German Mineral Resources Agency at the Federal Institute for Geosciences and Natural Resources (BGR) (Deutsche Rohstoffagentur)
DIW	German Institute for Economic Research (Deutsches Institut für Wirtschaftsforschung)
DLR	German Aerospace Center (Deutsches Zentrum für Luft- und Raumfahrt)
dmp	Depletion mid-point—point at which fifty percent of the reserves have been exploited from a mineral deposit
DOE	US Department of Energy
EASAC	European Academies Science Advisory Council
EGS	Enhanced Geothermal System, technical term for a deep geothermal facility for energy generation whereby the permeability of the rocks for the heat transfer medium (usually water) is improved by technical measures such as hydraulic stimulation (fracking)
EITI	Extractive Industries Transparency Initiative
EPA	US Environmental Protection Agency
EROI	Energy Return of Investment, or harvest factor, that relates the energy invested to that which is produced
ESYS	Energy systems of the future (Energiesysteme der Zukunft)
EU-25	25 member states of the European Union before the expansion in 2007
FAO	Food and Agriculture Organization of the United Nations
FCV	Fuel cell vehicle
FONA	Basic program of BMBF "Research for Sustainable Developments"
GDP	Gross domestic product
GEOMAR	GEOMAR Helmholtz Centre for Ocean Research Kiel
GFZ	GFZ German Research Centre for Geosciences Potsdam (Geoforschungszentrum)
GLR	Weighted country risk
GRI	Global Reporting Initiative
GtL	Gas to liquids includes the technologies that convert gas to liquids, for example natural gas to liquid fuel
HEV	Hybrid electric vehicle
HHI	Herfindahl-Hirschman Index
HIF	Helmholtz Institute Freiberg for Resource Technology (Helmholtz-Institut Freiberg für Ressourcentechnologie)
HWWI	Hamburg Global Economics Institute (Hamburgisches Weltwirtschaftsinstitut)
IAI	International Aluminum Institute
ICMM	International Council of Mining and Metals

IEA	International Energy Agency
IFC	International Finance Corporation
IGF	Intergovernmental Forum of Mining, Minerals, Metals and Sustainable Development
ILO	International Labor Organization
IMF	Remote Sensing Technology Institute (Institut für Methodik der Fernerkundung)
IPCC	Intergovernmental Panel on Climate Change
IRENA	International Renewable Energy Agency
ISMI	International Strategic Minerals Inventory
ITO	Indium tin oxide, indium oxide with tin additive is relevant to the production of thin sheet solar cells of LCD displays
IUPAC	International Union of Pure and Applied Chemistry
IZT	Institute for Futures Studies and Technology Assessment (Institut für Zukunftsstudien und Technologiebewertung)
JRC-IET	Joint Research Institute of the European Commission—Institute for Energy and Transport
LCD	Liquid crystal display
LED	Light-emitting diode
LNG	Liquefied natural gas
KfW	KfW—German state-owned promotional bank (Kreditanstalt für Wiederaufbau)
MB	Metal Bulletin
MEG	Metals Economics Group
MMSD	Mining, Minerals and Sustainable Development Project
NMMT	National Masterplan of Maritime Technologies (Nationaler Masterplan Maritime Technologien)
NPP	Net primary production
NRC	National Research Council of the National Academies of the USA
OECD	Organization for Economic Cooperation and Development
OPEC	Organization of the Petroleum Exporting Countries
PGE	Platinum group elements
PHEV	Plug-in hybrid electric vehicle
ProgRess	German Program for Resource Efficiency (Deutsches Ressourceneffizienzprogramm)
REO	Rare-earth oxides
SI	Social investment
SQUID	Superconducting quantum interference device
SX/EW	Solvent extraction/electrowinning, a process for extracting raw materials by chemical solution processes and electrolytic separation from the solution
TEM	Transient Electro-Magnetics
UKERC	UK Energy Research Center
UNEP	United Nations Environment Program

URR	Ultimate recoverable resource, theoretical term that includes the total of exploitable occurrences of a raw material on Earth
US-CBO	US Congressional Budget Office
USGS	US Geological Survey
VDI	Association of German Engineers (Verein Deutscher Ingenieure)
VEBA	Vereinigte Elektrizitäts- und Bergwerks AG (German energy company)
WEEE	Waste electrical and electronic equipment
WGI	World Governance Index
WING	Innovations of Materials for Industry and Society (Werkstoffinnovationen für Industrie und Gesellschaft)
WTO	World Trade Organization

Units

bbl	Barrel—used as a measure of volume for crude oil—is the measure for volume of crude oil products typically used in the hydrocarbon industry
Cu-eq	Copper equivalent
€	Euro
EJ	Exajoule, equivalent of 10^{18} J
gC/m^{2}yr	Gram carbon per square meter and year
g/t	Gram per tonne
Gtoe	Gigatonnes oil equivalent (1 toe = 41.868 MJ)
lb	Pound (unit of weight) (one lb = 0.45359237 kg)
IGK-$	International Geary-Khamis dollar, one IGK-$ is equivalent to the purchasing power of one USD normalized at a specific time
kg	Kilogram
kJ	Kilojoule
m^3	Cubic meter
MMcf/d	One million cubic feet per day (used as a measure of natural gas production, 1cf is 0.02832 m^3
Mt	Megatonne, one million tonnes
MW	Megawatt, one million watt
MW$_{el}$	Electrical megawatt, used as a measure of capacity of geothermal power stations
MWP	Megawatt peak, unit for the maximum capacity of a photovoltaic facility
t	Tonne
USD	US dollar

Chemical Element Symbols and Empirical Formulae

CH_4	Methane
CO_2	Carbon dioxide
$CO(NH_2)^2$	Urea
D_2	Molecular deuterium (deuterium is the heavy isotope of hydrogen)
D_2O	Heavy water
H_2	Molecular hydrogen
H_2O	Water
K_2O	Potassium oxide
N_2	Molecular nitrogen
NH_4^+	Ammonium
NO_x	Synonym for nitrogen oxide
NO_3^-	Nitrate
N_2O	Nitrous oxide
O_2	Molecular oxygen
PGM	Platinum group metals—including ruthenium, rhodium, palladium, osmium, iridium, and platinum
REE	Rare-earth elements
U_3O_8	Triuranium octoxide

Triuranium octoxide is an uranium (V,VI) oxide compound. It is primarily produced by processing of uranium to the marketable "Yellow Cake", and is used as the basic measure for the uranium content. Uranium can occur in compounds with the oxidation states from +2 to +6, but usually occurs in nature with the oxidation states +4 or +6.

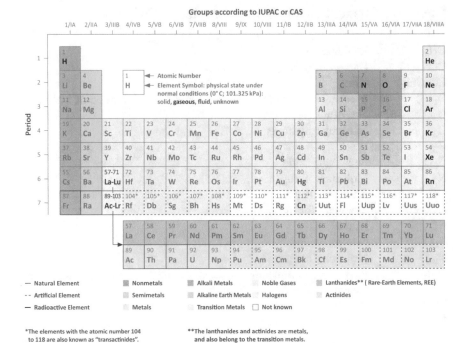

Groups according to IUPAC or CAS

— Natural Element   ■ Nonmetals   ■ Alkali Metals   Noble Gases   ■ Lanthanides** (Rare-Earth Elements, REE)

-- Artificial Element   ■ Semimetals   ■ Alkaline Earth Metals   Halogens   ■ Actinides

— Radioactive Element   Metals   Transition Metals   ☐ Not known

*The elements with the atomic number 104
to 118 are also known as "transactinides".

**The lanthanides and actinides are metals,
and also belong to the transition metals.

Summary

Three groups of natural resources are essentially required for future energy systems: fossil energy raw materials (gas, oil, coal), biomass, and mineral raw materials (particularly metals). Whereas the requirements for fossil raw materials are expected to decrease over the long term, and biomass can only cover a small proportion of the energy needs, the requirement for those metals that are required for the construction of renewable energy facilities and low carbon technologies will increase. The transition to new energy systems will therefore be accompanied by changes in the required raw materials. A reliable supply of the necessary raw materials is therefore an essential condition for successfully implementing the energy transition.

Germany—An Importing Country

The new energy technologies will require the same raw materials as used in many other advanced technology products. The energy sector will compete with the automobile industry as well as the electronic, IT, and communication sectors for the technology metals such as copper, cobalt, and platinum group metals, the special metals such as indium, tellurium, gallium, and germanium, or the rare-earth elements. Germany is, to a large extent, dependent on importing metals because it does not have its own metal mine production, and the recycling rates are very low for some raw materials such as the rare-earth elements, indium, tellurium, gallium, and germanium. These commodities are traded on the international markets, and therefore, the global demand is critical to their availability in Germany.

In contrast, the demand for construction raw materials, such as sand and gravel, and several industrial minerals, for example kaolin and gypsum, is covered by domestic production. The construction raw materials have, in principle, an unlimited availability but are in fact constrained by the competing uses for the land, such as for nature conservancy or the protection of drinking water.

A high proportion of the fossil fuel requirements such as gas, oil, and coal, is also imported. Only in the case of lignite, for which Germany is the world's biggest producer, does the domestic supply cover the demand. The proportion of fossil fuel requirements covered by imports is greater than two-thirds, and the import of energy raw materials is therefore of significant economic importance. Eighty percent of the demand for biomass fuel is covered by domestic production, but biomass for stock feed is now imported.

Increasing Global Demand

The raw material requirements for national economies change according to their advance in industrialization and economic development. The infrastructure and production facilities for the manufacturing industries must first be established. The material intensity, which is the ratio of the raw material requirements to the gross domestic product, sharply increases during this phase of development. At a specific level of development, the economy manages to produce more valuable products from the same quantities of raw materials, and the importance of the service sector increases. During this phase, the material intensity decreases, but even so the absolute quantities of the required raw materials can further increase. General predictions on the trends about the demand for raw materials in the future can be derived from this scenario: Up until the end of the previous century, 70–80% of the raw materials were used in the established industrial countries, but now China is the largest consumer of nearly all the important raw materials. For some of these raw materials, including steel, China has already passed the maximum material intensity point. It must be expected that in the long term the increase in demand for other raw materials will also slow down, but the timing is difficult to predict. From about 2020, additional populous countries with developing and emerging economies, such as India, Indonesia, or Brazil, are predicted to follow a growth pattern similar to that in China, however not as strong or rapid. The alignment of these countries to the levels of prosperity in the western industrial nations will result in further increases in the demand for raw materials to beyond 2050.

Geological Availability

Are the existing natural resources of raw materials sufficient to cover the increasing global demand? Several studies on natural resources address this question by applying the so-called static range of availability, which is the ratio of the reserves (known resources that are economically mineable) to the annual consumption. This quotient is often erroneously interpreted as the number of years for which there is enough material to satisfy demand, but this approach ignores that there are further resources and "geopotential" in addition to the known reserves. The additional resources are known, but are not economically mineable with current technology and at current prices. Geopotential includes those deposits that are not yet known, but can be reasonably expected to occur in specific geological terrains. Geopotential and resources can be converted to reserves by further exploration, development of mining, production and processing technologies, and increasing market prices. The reserves of most mineral raw materials therefore tend to increase during consumption, as is demonstrated by the example of oil, whereby the reserves have increased disproportionately to the consumption—from 1950 to 2013 the consumption increased eightfold, but the reserves increased by twenty times.

Most deposits of mineral raw materials would need to be classified in the field of geopotential, and despite increasing consumption, there is no geological reason to presume that there will be shortages that could threaten the implementation of the energy transition. The availability of these raw materials at competitive prices in the market is much more important. In general, the markets for raw materials can react flexibly toward changes in demand: Temporary shortages result in price signals that are then reflected by the supply and demand situation. Supply can be increased by exploration efforts as well as technical advances in mining, production and processing technologies so that resources and deposits with geopotential can be transitioned to reserves and production for the markets. In addition to the primary mine production, raw materials can also be recovered from end-of-life products and infrastructure (secondary deposits). These sources can be used as a measure to compensate for shortages in the supply chain. From the consumer's point of view, shortages and high prices encourage a more efficient and more economic use of the raw material, for example, by more efficient manufacturing systems or use of other technologies, or even substitution by another raw material. The interaction of this market mechanism is also known as the feedback control cycle of raw material supply, and, as a result of which, the average real prices for most mineral raw materials have not increased substantially during the last centuries.

Price Peaks Related to Sudden Increase in Demand

Short-term price peaks repeatedly occur despite the long-term trend noted above. This is because there is usually a time lag before the supply can cover these sudden increases in demand. On the one hand, it usually takes about 10 years from the date of discovery of a mineral deposit to develop a mine, and similarly, increasing the capacity at an active mine site also requires time to make the necessary expansions. Since there is a global market for nearly all raw material commodities, even small increases in demand by a few percent can result in significant increases in price and therefore affect the short-term supply situation. On the other hand, the geographic distribution of the mineral occurrences and the structure of the mining business is leading to a concentration of the industry, so that the trend is toward more mineral resources belonging to fewer companies in fewer producing countries. This is resulting in a distinct tendency toward the formation of oligopolies. Individual companies or countries can thus, for example, restrict production or exports to influence the commodity markets, which may lead to a consequent deterioration in the security of supply.

Manufacturing industries that require raw materials must be able to evaluate the risks along the supply chain. Businesses can develop alternative strategies only if potentially critical raw materials are identified in advance, and these can include diversification of the supply sources, increased use of domestic secondary materials, or develop the possibilities for substitution and more efficient use of materials in their production processes. These companies can mitigate the supply risk by establishing consumer cooperatives, long-term supply contracts with price escalation clauses, and appropriate hedging measures.

The information required for assessing these risks is provided in Germany, for example, by the German Mineral Resources Agency (DERA) at the Federal Institute for Geosciences and Natural Resources (BGR). The companies could be supported politically in their efforts to ensure secure supply of required raw materials by, for example, the European Union and World Trade Organization (WTO) reducing the trade barriers in global commodity markets. However, these measures would normally come into effect over the medium to long term.

Evaluation of Criticality

The criticality of raw materials can be evaluated according to various parameters. An important early warning indicator is the ratio of reserves and annual consumption. As already noted, the continuous growth in reserves means that this is not a fixed figure but a snapshot in a dynamic system. However, impending supply bottlenecks can be identified by long-term surveillance of this ratio: If the value falls below 10–15 years, which is the range of typical development times for mining projects, then the raw material could be classified as potentially critical. Until now, this is only applicable to antimony and tin, which are two metallic raw materials that are not critical to future energy systems.

Another important indicator is the weighted country risk, which evaluates the conditions in the producer countries such as political stability, corruption monitoring, and effectivity of the public sector. The resulting risk rating is then weighted according to the proportion of raw material production from that country to the overall global production. Raw materials are critical if they are primarily sourced from relatively few producer countries, all with high country risk ratings. While the static range of availability is primarily an indicator for the required exploration activities, the weighted country risk is an indicator of the political issues required for the dependable supply of the raw material. In addition to the weighted country risk, the evaluation of the criticality should also include an assessment of the potential cost to the economy in case of the raw material being unavailable. If the possibilities for substitution or recycling are low, then the risk is increased.

Tabulations or lists of raw material risk, such as those issued by the European Commission in 2010, 2014, and 2017, can assist industry to prepare for possible bottlenecks in the supply of specific raw materials and to develop appropriate alternative strategies. German companies often use intermediate products that are higher in the value-added chain as their principal input material, and this must be taken into consideration in the assessment of the supply situation, but there is very little relevant information. The DERA has therefore begun to include important intermediate products in its criticality analyses. Further to this detailed analyses have been issued for copper, tin, zinc, zircon, antimony, and tungsten, the platinum group metals, bismuth and lithium. It is notable that the concentration of supply for numerous mineral raw materials and intermediate products is much greater than that for fossil fuels, such as crude oil. The development of renewable energy technologies, within the scope of the energy transition, could therefore lead to new trade disputes.

Scenarios that assess technology developments and demand-side trends are very important for evaluating the future raw material requirements. However, changes in demand are increasingly difficult to predict because of the shorter production cycles. This is clearly demonstrated by the television manufacturing industry: The transition from tubes to LCD television screens materialized within only 2 years and, instead of barium and strontium, indium and tin became the relevant metallic raw materials.

Metals for the Energy Transition

All the relevant technologies must be taken into consideration for an assessment of the raw material requirements during the energy transition. Furthermore, an energy system that is primarily based on decentralized technologies for the exploitation of renewable energy is more diverse than the current system. Wind power and photovoltaic facilities, various battery systems, hydrogen storage, electric vehicles, and light-emitting diodes (LEDs) are just a few examples. It is estimated that about 45 different technologies will be important for the development of the energy system.

Critical raw materials mainly include the rare-earth elements, the platinum group metals, cobalt, as well as indium, tellurium, and other rare metals. As much as 95% of the mine production of rare-earth elements, which are required for energy-saving lamps and permanent magnetics in wind energy plants, motors, and generators, is in China. In the future, these rare-earth elements may also be increasingly required for batteries and photovoltaic facilities. The platinum group metals are particularly important for fuel cells and hydrogen electrolysis—and therefore for the technologies that are potentially crucial for the energy transition such as long-term storage and power-to-gas conversion. Cobalt is an essential functional metal in lithium-ion batteries which are key element for the expected strong increase of electric vehicles.

In addition, there are several raw materials for which different opinions on their criticality have been expressed in various commodity studies. These include nickel, niobium, tungsten, gallium, germanium, selenium, vanadium, silver, graphite, rhenium, and hafnium. Depending on the assumptions pertaining to the future technological developments and the trends in the commodity markets, these raw materials might also be regarded as critical.

Many critical, or almost critical, metals—including indium and tellurium—are produced as by-products in mines for other metals. The feedback control cycle of mineral supply does not fully apply for the by-product metals. For example, indium is produced as a by-product in zinc mines, and in case of a shortage of indium, the feedback control cycle of mineral supply is only effective to a limited extent because the mine operator is unlikely to increase production capacity only for indium. There are often only a few producers and customers, and the market is less transparent than for commodities that are traded on the metal exchanges. The future availability of the by-product elements, therefore, is difficult to assess.

New Deposits at Depth and on the Sea Floor

Research into mineral deposits decreased throughout the world during the long period of over-supply of natural resources in the 1980s, but this situation reversed and global exploration increased with the trend to higher prices from 2005 to 2012. With the following decrease of metal prices also exploration activities declined, but picking up again since 2017. The best chances for discovering new deposits are at depth. Near-surface ore deposits have been explored and mined out, but since the 1990s more new deposits have been discovered at depth. The discovery technologies for marine raw materials are already advanced, although the technologies for the production and processing of raw materials from the deep sea, such as polymetallic manganese nodules, are still in the early stages of development.

Nonetheless, the importance of these marine resources will probably increase in the future. The Federal Government has therefore acquired concessions in the Pacific for polymetallic manganese nodules, which contain copper, nickel, and cobalt, as well as a concession for base metal containing massive sulfide mineralization in the Indian Ocean. The massive sulfides also contain by-product elements, including several important "high-technology" elements.

Recycling of Important Metals

The secondary deposits, which are those raw materials in end-of-life products and infrastructure such as vehicles, computers, buildings, electrical cables, and waste dump materials, represent a substantial potential in Germany. If these secondary deposits are effectively exploited, they can add significantly to the resource base of raw materials in the national economy. In principle, by applying appropriate metallurgical processes, it is possible to recover metals from secondary materials of the same quality as that obtained by processing materials from primary sources (aluminum being one exception). Recycling can therefore contribute to reducing the dependence on the supply of critical raw materials from primary mineral deposits. Furthermore, the development time and capital requirements for recycling projects are often less than for primary mineral deposits, and recycling enjoys an enhanced social acceptance as compared to that for mining.

Recycling of pure metals requires less energy as compared to recovering metal from primary deposits. However, a high recycling ratio would require that an increased amount of material must be recovered from those secondary deposits with a lower metal content and more complex composition, and this in turn would require more energy. Since the primary deposits will themselves become increasingly complex, the proportion recoverable by recycling with an optimal use of energy will increase.

High recycling rates have so far only been achieved for the major and precious metals. However, the recycling rates for the rare-earth elements and high-technology metals such as indium, germanium, gallium, or tellurium remain unsatisfactorily low because the metallurgical infrastructure is insufficient for the recovery of these elements, which are often finely distributed. The ratio between the actual production from the secondary deposits and their potential depends on the efficiency of the whole of the processing chain: collection, disassembling, preprocessing, and metallurgical processing. The waste legislation and its implementation are also an important component in this chain. For many of the metals important for future energy systems, excessive losses already occur during the first stage of collection. Even today, only a small proportion of normal consumer electronic devices find their way to an efficient recycling plant. Even though the content of metals per tonne of electronic scrap is often higher as in a tonne of primary ore, the specific metal value for individual consumer products is low, for example approximately one euro for a mobile telephone. There is therefore an inadequate economic incentive for the consumer to bring the product to a recycling plant.

The global recycling economy suffers a significant loss of metals because of illegal and dubious exports of electronic scrap and other end-of-life products in regions with inadequate recycling infrastructure. The processing of end-of-life appliances in plants outside the European region, which are less efficient and operate at lower standards, is often economically more attractive although often also related to serious consequences for health and environment, and with significantly lower metal recoveries as is possible in modern industrial plants.

The quality of the secondary materials, and therefore the recycling rate as well as the energy demand for the recycling, is closely related to the design of the products and appliances. For example, if the components containing valuable metals—such as magnets, batteries, or electronic components in vehicles and electrical appliances—are easily accessible, then they can be extracted before the shredding process and the metals can be recovered more efficiently.

Although increasing quantities of raw materials from geological deposits are used in infrastructure and other products, and will therefore finally enrich the secondary deposits, a 100% recycling rate is, in the long term, neither realistic with respect to the energy requirements nor would it cover the total demand. Furthermore, metals can only be recovered at the end-of-life stage of any product. Therefore, the resources of recoverable raw materials required for a rapid expansion of new technologies will initially be low, presuming that the relevant materials are not present in sufficient quantities in other end-of-life products. During the conversion of the energy systems, the proportion that can be met by production from secondary sources therefore also depends on how quickly the development of new technology systems advances and the lifespan of the products.

Water and Energy Requirements of Mining

In the mining industry, water is primarily required in the processing of ores. Because many mining districts are located in arid or semiarid regions, some experts consider that water availability will be a limiting factor for future raw material supplies. The requirement for freshwater can be partly offset by using brackish or salty water, and desalinated seawater is also used in mining, but this requires a significant increase in the energy input.

The energy input per tonne metal will probably increase with time as increasingly deeper ore deposits must be exploited and more complex ores are processed. This is obviously relevant to the environmental balance. Even today, 8% of both the global energy demand and the CO_2 emissions are directly attributable to mining.

For energy raw materials, the proportion of energy that must be expended for the development, production, and deployment of an energy production plant is analyzed by the so-called energy amortization calculation, and is measured as Energy Return of Investment (EROI). This value is very strongly dependent on the conditions of the deposit, and, for example, for natural gas it varies between 15 and 200. The amount of energy required for exploiting those raw materials from which the plant is constructed is also important for renewable energy. The EROI values for photovoltaic and wind energy plants are significantly higher than for most of the bioenergy scenarios.

If the energy is derived from renewable sources, then the energy requirements for the development of resources are obviously less problematic with respect to climate change. The energy balance of mining companies can be improved if the production is customized to the changeable supply of electricity from solar and wind sources. It is conceivable that companies will deliberately use inexpensive surplus solar and wind energy to process ores with a particularly low metal content. Even today, many open-pit operations have two ore stockpiles for high-grade and low-grade ores, and the latter is processed at times with favorable, namely high, commodity prices.

Social Acceptance—A Necessary Condition

Even if all the technical and economic requirements for a successful mining project are fulfilled, there is one further important requirement: social acceptance. The local communities must accept and support the mine, or at least tolerate it. The gaining and maintenance of social acceptance for mineral resource exploitation, the so-called Social License to Operate, are becoming an increasingly significant challenge for the mining industry.

The proportion contributed by mining to the gross domestic product is continuously decreasing in the industrial nations such as Germany, France, and Britain, all of which were producers of mineral raw materials in the past. As the importance of the mining industry decreases, so also does the public knowledge about the significance of natural resources and their interest in raw material-related topics declines. Mining is often coupled with a negative image and is associated with environmental damage and dangerous working conditions for the miners. This is also now increasingly the case in the classical mining countries such as Canada and Australia, where exploration concessions are prohibited over a significant proportion of their surface area. Resistance to mining projects is also becoming an issue in the resource-rich emerging and developing countries, which are often dependent on their resource exports. However, this is likewise a reflection of the legacy from historical mining in many countries, such as Bolivia, Chile, and Peru, including fragmented social structures and contaminated sites. Large mineral deposits do not only bring wealth and economic growth. In many situations in emerging and developing countries, mining has resulted in islands of economic activity that preserve or even enhance social inequality in the community, rather than contributing to the overall development of a region or country. Furthermore, indigenous people in many resource-rich countries are impacted by the consequences of the mining activities although they themselves have no use for these raw materials.

The conflicts of interest associated with mining are often extremely complex and vary from country to country. There are several factors that influence the acceptance or rejection of mining by the public in any one country, including: the level of development in the country and economic dependency on exploitation of raw materials; on foreign currency and tax income; jobs and infrastructure development that is generated by mining; and on the local environmental impact caused by the mining and smelting. Sustainable and socially and ecologically acceptable exploitation of natural mineral resources can only be established after the various interests have been discussed and evaluated. Good governance structures are an important element: Export-oriented mineral production will only result in an overall positive impact on the country if effective administration is established together with political, economic, and public interests, so that collective tasks can be managed for the public good.

Social acceptance for resource exploitation can only be achieved if the public is convinced that their values are respected, the environmental impact is minimized, and the economic advantages are reflected by jobs and improved infrastructure.

Open-pit mining projects are often the most strongly criticized. The impact on the landscape is very much more than that for an underground mining project, and in some cases, whole villages must be resettled. Furthermore, the subsequent consequences from the mining such as the water balance are difficult to estimate. The question must therefore be asked: Is the currently observed trend in mine planning to convert, due to lower costs, underground mines into increasingly large open pits a move in the wrong direction over the long term?

It is interesting to note that in Austria open pits are transitioning to underground operations due to environmental constraints. These new endeavors to constrain the impact of resource exploitation on the environment and landscape can foster, and even enhance, the necessary social acceptance. This trend would be sustained by the development of more efficient underground machinery.

Environmental and Social Standards

Inadequate environmental and social standards are a cause not only for health and environmental hazards, but also result in a distortion of the competition on the commodity markets because the social and environmental costs are thereby externalized and transferred to the public purse. The necessary improvements of environmental and social standards are a major, if not the most important, challenge to be addressed internationally by the mining industry.

Because of various serious incidents, the major international mining companies have cooperated on initiatives and commitments to social and environmental standards. One example is the International Council of Mining and Metals (ICMM) that sets the standards for about 30–40% of the global resource exploitation. However, there is no member from China, although China is the world's largest mining country. Regular monitoring, evaluation procedures, and education activities can assist in improving the standards, and these are undertaken by, for example, the Global Reporting Initiative (GRI), which is an independent international organization that was founded in 1997 with the participation of the United Nations Environmental Program (UNEP). Unfortunately, one problem is that some of the medium- and small-scale mining companies do not abide by the standards of these initiatives. Although these companies only contribute a small proportion of the global production, they often cause a disproportionately high environmental impact as compared to their production.

The international banks also have an important role in the enforcement of these standards because they can ensure that the maintenance of standards is a condition of the mine financing package. Large-scale international mining projects are commonly financed to about one-third by the company's own capital and two-thirds with capital raised from banking consortia. The international banks and private investors therefore have to accept a definite responsibility for the social and environmental standards.

Fossil Fuel Resources

More than 80% of the global primary energy consumption is currently provided by the fossil fuels—oil, gas, and coal. The International Energy Agency (IEA) presumes a continuous increase in energy consumption until at least 2040, whereby the growth is focused not in Western Europe but will primarily take place in China as well as various emerging countries. Fossil fuels also provide 80 percent of the energy demand in Germany, although the consumption has decreased slightly since 1990; and until 2050, the primary energy consumption should have been halved as compared to 2008.

Even if the electricity production from wind power and photovoltaic continues to be quickly expanded, coal or gas (either natural gas or biogas) power stations must continue to be available for the foreseeable future—at least for as long as there is no long-term electricity storage capability. Coal will be available in the long term and is relatively inexpensive. However, as compared to all the other fuels, coal causes the most CO_2 emissions. Gas power stations, on the other hand, have the advantage of being more flexible as compared to coal power stations and can therefore better compensate for power fluctuations related to the feed from wind power and photovoltaic generators. Furthermore, the combustion of natural and biogas is significantly cleaner than that of coal, albeit more expensive.

The resources and reserves of hard coal, lignite, and uranium are so large that no restrictions on supply are to be expected even if the demand continues to increase. However, since the decision by Germany to withdraw from nuclear energy, uranium will no longer be relevant for supplying the German energy market.

Lignite is almost entirely sourced from domestic production, but the proportion of domestic hard coal has drastically decreased in the past 10 years, and in 2013 was only 13%. The last German hard coal mine will close in 2018 due to the termination of subsidies for the domestic hard coal production.

As compared to coal, oil and natural gas are scarce natural resources. In 2013, natural gas from German production covered about 12% of the domestic consumption, and for domestic oil production, it was only 2%.

Oil is primarily imported from Russia, Norway, the UK and from politically less stable regions such as the Middle East and North Africa. Oil is the most expensive energy resource and is characterized by frequent, short term, price increases. From 2000 to 2008, for example, the price of oil increased more than ten times. These price increases are more often the result of political events, such as the earlier restrictive measures agreed by OPEC, the Kuwait crisis, or the Iraq war, rather than a reflection of increased production costs.

Germany is also dependent on a few countries for its natural gas. While some of the natural gas is derived from the Netherlands, this source will decline due to the decreasing reserves. As a result, the dependency on imports from Russia will increase, but this can be offset by more imports of liquid natural gas (LNG), which is delivered by tankers.

The exploitation of oil and natural gas from unconventional deposits in Germany, which are hosted by very dense and compact rocks (shale gas and shale oil, for example), presents an alternative scenario, but the production is only possible by using additional technical measures. The best-known technology is hydraulic stimulation by fracturing (fracking), which has been the subject of controversial public discussion. The development of unconventional oil and gas resources in the USA demonstrates how technical innovation can radically impact the resource supply. It is very possible that the USA will become self-sufficient in natural gas, and even the current moderate prices for oil and natural gas on the world markets are primarily related to the increased production of shale gas and shale oil.

Coal seams are another potential source for the production of natural gas. This might be possible in Germany, particularly in northern North Rhine-Westphalia, but further investigations are required. In the long term, it is even possible that the methane clathrates on the ocean floors could become relevant as a source, since these occurrences are considered to be very large even if the exact volumes and production costs have not yet been quantified. The production is currently not economic, although there is land-based production from occurrences of gas clathrates in the permafrost regions.

Even if the production from unconventional deposits should significantly increase, the resources of natural gas and oil will remain limited. Despite the current period of moderate prices, it is considered that prices for oil and natural gas will increase in the long term due to higher demand.

Bioenergy

Approximately, 10% of the global demand for primary energy is sourced from biomass, and in many developing countries biomass, in particular wood, is the principal source of energy. In contrast in 2014, biomass provided about 7.5% of the energy requirement in Germany of which 50% was for heating, 25% for electricity generation, and 16% was used as fuel. About 20% of the bioenergy was imported.

As compared to photovoltaic and wind power facilities, bioenergy has a relatively low spatial efficiency. The greenhouse gas balance is also less favorable, and the price for saving one tonne of CO_2 is generally higher. However, because it is storable and has a high energy density, bioenergy can perform some of those functions in the energy system for which wind and solar energy are less suitable. For example, it can be used to bridge the electricity requirements during longer periods of calm winds or be used as control energy. Bioenergy can also replace the use of fossil fuels in transport, in particular heavy transport (goods vehicles and freight ships).

The estimates of the amount of bioenergy from agricultural biomass that will be available globally in 2050 are wide apart. They range from 50 EJ (exajoules) per annum (current status) to 500 EJ per annum. The different figures arise from the diverse assumptions for, among others, the expected increases in the harvest yields, the amount of agricultural land required for food, and the sustainable quantities of available water.

The evaluation of greenhouse gas emissions from agriculture and other environmental impacts has an important effect on the potential of bioenergy. Arable land, meadows, and pastures must be farmed, fertilized, and sometimes also irrigated in order to provide high yields for biomass. Only forest land does not usually receive additional mineral fertilizers. This all has consequences for the greenhouse gas balance: It is generally neutral only for the sustainably cultivated forests. In contrast, the intensive cultivation of arable land, meadows, and pastures results in the release of carbon dioxide, methane, and nitrous oxide as greenhouse gases. In 2011, nearly 8% of the greenhouse gas emissions in Germany were derived from agriculture, and the global proportion is as high as about 20%. Even if, as compared to fossil energy fuels, the release of greenhouse gas is less per energy unit, and the use of agricultural biomass as an energy source can by no means be regarded as climate neutral.

Other environmental impacts related to intensive agriculture include the decrease in biodiversity, very significant requirements of water, and contamination of water sources by excessive use of fertilizers. The quality of the soils can also deteriorate. Despite these environmental risks, the total area under cultivation for intensive agricultural production is expected to increase by about 5% between 2005 and 2050.

Globally, the quality of the soils and the availability of water for agriculture have a limiting effect. In several parts of the world, soil is currently being lost at a rate one hundred times quicker than it is being formed for various reasons, including erosion, salinization by salt from evaporating irrigation water, compressed by heavy machinery, loss of soil carbon that is oxidized to CO_2, and sealing by excessive construction.

Irrigation is very important for the agricultural productivity in many regions, and it currently accounts for about 70% of the global use of fresh water. Salt water is available in virtually unlimited quantities and can be converted to fresh water by the energy-intensive desalination process. The water issue is therefore becoming an issue of energy availability.

Because of these issues, it is unlikely that there will be a significant increase in the availability of bioenergy. Currently, bioenergy contributes about 10% of the global energy consumption, and this proportion is therefore likely to decrease as global energy consumption increases. In comparison to mineral resources, where production can be increased according to the feedback control cycle of mineral supply, the situation for biomass is more tightly constrained within narrow limits.

As a result, the measures to secure supply are logically focused on the demand, so that biomass in all sectors is used as efficiently as possible and bioenergy is only used in situations that provide the greatest advantage to the total energy system.

The requirement for agricultural biomass can be reduced by a more efficient supply chain of foodstuffs and by nutrition with less animal products. The latter also requires large land areas, and so newly available agricultural lands could then be used to produce bioenergy. The use of agricultural biomass waste also has significant potential as an energy source.

Fertilizers

The fertilizers for agriculture include nitrogen fertilizers, phosphates, and potash fertilizers, and there are no substitutes for these materials. Nitrogen fertilizers can be produced from atmospheric nitrogen by the Haber-Bosch process in virtually unlimited quantities. However, constraints in the future could include the greenhouse gas emissions that result from the energy-intensive production process, and the environmental impacts derived from fertilizers.

There is unlikely to be a resource scarcity for potassium. Apart from geological deposits, potassium is present in virtually unlimited quantities in the seawater, from which it can be recovered by evaporation.

German agriculture currently uses about 650,000 tonnes phosphate per year. The ratio of reserves to mine production is currently estimated to be about 300 years, so that this early warning indicator does not suggest an impending scarcity in the supply of phosphate. However, in contrast to nitrogen and potassium, the available resources of phosphorous are not unlimited, and phosphorous could be regarded as a sort of critical raw material. There are new international approaches to identify the availability based on the geological potential, and observe the trends. Recycling of phosphate from sewage sludge is in part technically feasible, but is not currently economic. About half of the phosphate requirements for German agriculture must be imported, and the other half is provided by liquid manure.

Conclusion

The global commodity markets have until now always reacted flexibly to shortages of specific raw materials, so that supply bottlenecks and price spikes have been short-term events. This will also remain the case in the future so that the implementation of the energy transition will most likely not fail as a result of insufficient availability of raw materials. However, trade conflict situations could increase because of the high concentration of sources for many of the raw materials required for the energy transition.

Good international trade relations as well as innovations both in mining and recycling are necessary to ensure a sustainable and reliable supply to the industry of the necessary raw materials at competitive prices.

Research and development in the fields of mineral exploration, mining, and metallurgical processing technologies could contribute to limiting the prices for primary raw materials to reasonable levels, despite the increasingly difficult and challenging settings of mineral deposits.

Substantial development in the fields of recycling, material efficiency, and substitution for the raw materials of strategic economic importance reduces the dependency on imports, the energy requirements, and the environmental impacts from primary production. These technical advances should be accompanied by appropriate measures such as support for global free trade, avoidance of monopolies, purposeful cooperation agreements with resource-rich countries, a concerted approach to international resource policies, and legislation in support of recycling and a more circular economy, as well as new forms of stakeholder collaborations along product value chains.

The elimination of trade restrictions must not be allowed to occur to the detriment of environmentally and socially acceptable production of primary mineral resources. Establishing and maintaining the highest environmental and social standards are fundamental conditions for the social acceptance of resource exploitation. At the same time, a truly sustainable energy transition can only be achieved if the stakeholder along the raw materials value chain, up to and including the end user, insists that such standards are globally maintained. The key question is not about the sufficiency of the raw materials for the energy transition, but about their long-term availability and the footprint linked to their production. The advantages of green energy can be significantly reduced by environmental and social impacts in the upstream sectors of the supply chain.

Chapter 1
Introduction

The worst impacts from anthropogenic climate change can probably only be avoided if mankind reduces the emission of carbon dioxide as quickly as possible, and alternatives are developed to the combustion of crude oil, natural gas and coal. Although there are today alternative technologies already available, there is still a long way before energy generation is climate neutral. Although renewable energy supplied about 13% in 2014 and nearly 15% in 2016 of Germany's requirements (gross primary energy) ([1] (Umweltbundesamt), [2]) the proportion of modern renewable energy (excluding biomass that is traditionally used in developing countries for cooking and heating) is about 10% of the global energy consumption [3]. Currently oil, natural gas and coal (and nuclear energy) therefore remain the most important suppliers for generating energy.

Germany and other countries are thus pushing ahead with the expansion of renewable energies. China is a particularly impressive example, as just a few years ago there were scarcely any wind energy facilities in the country, but now it is has become the frontrunner in developing wind energy. By the end of 2014 the nominal output from their wind energy facilities was almost as much as from all the European facilities together. This example emphasizes how fast the expansion of renewable energy can be, as well as the scale that it might become in the coming years. At the same time, other possibilities with new energy technologies will probably be developed, for example the markets for electrical mobility or storage for electricity derived from wind and photo-voltaic.

As the importance of the new energy technologies increases, so will the demand for raw materials also change with the same pace. While the dependency on imports of oil and natural gas will decrease, the expansion of the new infrastructure will require numerous new chemical elements, particularly various metals that were previously less in demand. The issue that therefore arises is if or how the demand for raw materials can in the future be met, so that the expansion of innovative energy technologies can proceed on a large scale.

© Springer International Publishing AG, part of Springer Nature 2019
F.-W. Wellmer et al., *Raw Materials for Future Energy Supply*,
https://doi.org/10.1007/978-3-319-91229-5_1

1.1   Will Raw Materials Ever Be Depleted?

At the beginning of the 1970s the question about the finite nature of resources became a public issue. The publication of the report *Limits to Growth* [4] by the Club of Rome was an important factor in this debate. The people in Germany and other western industrial countries were also faced with the effects of the oil price crisis, which spectacularly demonstrated how dependent the industrial nations are on the supply of raw materials to the global markets.

Since that time there have been numerous speculations about the dependability of the raw materials supplies in the future. These discussions are accompanied from time to time by impressive price spikes of specific commodities, such as most recently the rare-earth elements. Export restrictions and price speculation caused the price of these elements to rise very fast during 2009, by as much as 100 times in the extreme cases. The rare-earth elements are required for the manufacture of permanent magnets in computer hard-disk-drives or wind power generators, as well as rechargeable batteries and LED light diodes. This example demonstrates how new energy technologies can result in new dependencies, which can be just as problematic as the dependency on crude oil.

1.2   Objectivity in the Discussion

This analysis should contribute to increased objectivity in the discussions about the availability of natural resources. The so-called feedback control cycle of resource supply, which depicts the relationship between supply and demand on the resource markets, plays an important role in the discussion. The efforts required to advance exploration and develop new mineral occurrences are also explained.

At the same time, the analysis will also clarify why there should be no bottlenecks in the future, at least with respect to the geological resources of mineral raw materials. However, geological availability does not automatically imply a market availability. It is therefore very important for importing countries such as Germany to develop suitable strategies for ensuring a sufficient supply of the necessary raw materials from the world market, and to counter any possible price controls. This text, therefore, should provide an objective analysis of the global resource situation.

The raw materials important for energy generation are the focus of this discussion. Apart from the fossil energy fuels (crude oil, natural gas and coal) and biomass, these raw materials are primarily the mineral resources (in particular metals). One interesting point that should be noted is that metals are not consumed, in the exact meaning of the word, but are only used to manufacture products. Geologists therefore discriminate between the natural occurrences in the earth, or the geosphere, and those resources contained in today's products, infrastructure or in mankind's own space for living, which is known as the "technosphere". The question of how far an optimal recycling can contribute to an improved supply of raw materials will therefore be analyzed in detail.

1.3 Social Acceptance: A Condition for the Exploitation of Natural Resources

The geological availability and the technical and economic feasibility are, however, not the only conditions for exploiting natural resources. A "Social License to Operate" is an additional requirement, which means that the exploitation of resources must be supported, or at least tolerated, by the public. Because of the grievous mistakes during the past decades there are strong reservations about mining, and many people associate mining with the attributes "dark, dirty, dangerous". The improvement of the environmental and social standards should therefore be a major, if not the single most important, future task for the natural resource industries, and thereby garner a greater level of acceptance for mining among the population. This is particularly important for the supply of raw materials to the future energy systems. This analysis, therefore, should also provide a critical survey of the global exploitation of raw materials.

References

(Note: All Web links listed were active as of the access date but may no longer be available.)

1. Umweltbundesamt: *Anteil erneuerbarer Energien am Energieverbrauch*, Umweltbundesamt 2015. URL: http://www.umweltbundesamt.de/daten/energiebereitstellung-verbrauch/anteil-ern euerbarer-energien-am-energieverbrauch [accessed: 15.12.2015].
2. Bundesministerium für Wirtschaft und Energie: Erneuerbare Energien in Zahlen, Berlin: BMWi 2017 URL: https://www.bmwi.de/Redaktion/DE/Publikationen/Energie/erneuerbare-energien-in-zahlen-2016.pdf?__blob=publicationFile&v=8 [accessed 29.1.2018].
3. Renewable Energy Policy Network for the 21st Century (REN21): *Renewables 2015 Global Status Report*, Paris: REN21 Secretariat 2015. URL: http://www.ren21.net/wp-content/upload s/2015/07/REN12-GSR2015_Onlinebook_low1.pdf [accessed: 15.12.2015].
4. Meadows, D. H./Meadows, D. L./Randers, J./Behrens III, W. W.: *The Limits to Growth*, A Report for the Club of Rome's Project on the Predicament of Mankind, Universe Books: New York 1972.

Chapter 2
Fundamentals

2.1 Conventional Classification of Natural Resources and Definitions

Natural resources are generally subdivided into renewable and non-renewable raw materials (Fig. 2.1). Biomass products, such as wood, cellulose and starches, are the principal **renewable raw materials**. Renewable energy also includes the intangible wind, solar and geothermal sources. The **non-renewable raw materials** are subdivided further into two major sub-groups: energy raw materials and non-energy raw materials. The energy raw materials include the fossil energy sources such as coal, oil, natural gas and the radioactive elements uranium, plutonium and thorium that are important for nuclear energy. The metallic and non-metallic raw materials comprise the **non-energy raw materials**.

Because of their recyclability, metals are only used but not consumed. They can therefore almost be regarded as renewable raw materials. The non-metallic raw materials are subdivided into three additional sub-groups. The first group includes the bulk natural resources such as construction materials (sand and gravel) and the primary natural resources required for cement production (limestone, clay and marl). The second group includes the salts such as potassium salts. The third group is comprised of the industrial minerals such as phosphate and kaolin.

The industrial minerals include a variety of special raw materials for different applications that often require very precise chemical or physical specifications as well as combinations of different minerals. Water is also included as a non-metallic resource, and is differentiated into fresh water that has a limited availability, and salt water with virtually unlimited availability. Soil is also a distinctive raw material that, as the substrate base for the growth of plants, is of decisive importance to produce biomass.

© Springer International Publishing AG, part of Springer Nature 2019
F.-W. Wellmer et al., *Raw Materials for Future Energy Supply*,
https://doi.org/10.1007/978-3-319-91229-5_2

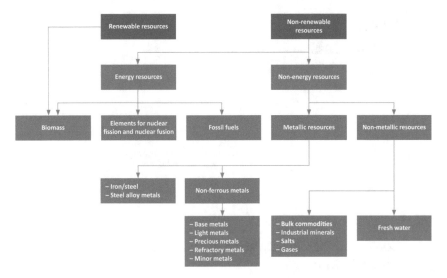

Fig. 2.1 Classification of Raw Materials (Authors' Diagram). Intangible renewable energy sources such as wind and solar energy or geothermal are not included

Those chemical elements that occur as fluids or gases in normal circumstances, or standard conditions,[1] are a special group of non-energy raw materials, and includes the elements mercury, bromine and helium. The noble gas helium is particularly important for future energy systems, and is used in large quantities for the cooling of technical facilities—in so-called refrigeration processes such as cryogenic systems [1] (see Sect. 5.4.2).

The metallic raw materials (refer to Appendix A, Fig. A.1) are essential to the future energy systems, and they are therefore discussed in greater detail in this analysis. The variety of metallic raw materials is immense. Some metals, for example iron, are produced in great quantities, such that the annual global production of iron and steel is about 1.5 billion tonnes per year. In contrast, the annual production of some electronic metals such as gallium or germanium is only a few hundred tonnes. The metals are normally subdivided into the following groups:

- Iron/steel and steel-alloy metals (for example nickel and molybdenum)
- Non-ferrous metals (base- and light-metals)

  ◦ Base metals (for example copper or zinc)
  ◦ Light metals (for example lithium, aluminum and magnesium)
  ◦ Precious metals (gold, silver and the platinum group elements or metals platinum, palladium, ruthenium, rhodium, iridium and osmium)
  ◦ Refractory metals (for example tantalum or tungsten)

---

[1]Reference conditions in chemistry are defined for the material specifications for each element: Standard Conditions are room temperature (25 °C or 298.14 K) and a pressure of 101.3 kPa. The reference temperature for Normal Conditions is 0 °C or 273.15 K.

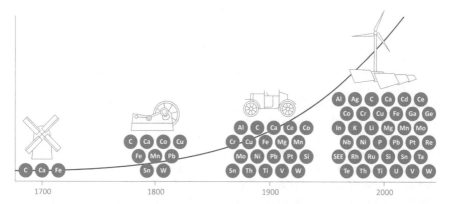

Fig. 2.2 The technology-related application of various elements with time. The number of chemical elements required for the energy technologies, such as the steam engine, internal combustion engine for motor vehicles, or the modern solar technologies, has increased significantly during the past 300 years. The increase in the number of elements required has increased almost exponentially since the beginning of the industrial revolution [*Source* Achzet et al. [26, p. 6], with permission of the authors (The element abbreviations can be taken from the Periodic Table at the beginning of the text)]

- ◦ Minor metals (for example antimony)
- ◦ Rare earth group of elements and electronic metals or semi-conductor elements (for example indium and germanium).

Rare-earth elements, the electronic metals and semi-conductor elements are also known as specialty metals. The term metalloid, which is normally used in physics and chemistry for silicon, germanium, arsenic, selenium, antimony and tellurium, is not commonly applied in the commodity business.

The boundaries between these categories are not clearly defined, particularly so for the distinction of the minor metals from the other groups of metals. As a result, the commodity business identifies technology metals as a group that includes, for example, the precious metals and other metals such as indium, gallium, germanium, antimony, selenium, silicon and tellurium.

Apart from most of the precious metals, many of the technology metals have only been used industrially during the few last decades (Fig. 2.2). Many of these metals are only produced, traded and used in small quantities as compared to iron [2]. Nonetheless, these metals are very important in, for example, the production of semi-conductors.

2.2 Reserves, Resources and Geopotential

With respect to exploitation of mineral raw materials, most experts define different classes of mineral deposits—the reserves, resources and geopotential. **Reserves** can be applied to those mineral deposits that have been identified by sufficient sampling and can be exploited economically with the currently available technology. A mineral deposit is identified as a **Resource** if it is already known and understood (certainly at different levels of knowledge), but cannot be economically exploited with current technologies or at the current prices. Commodity price levels that do not cover the extraction costs, lack of suitable infrastructure, insufficient grade of the mineralization, or technological problems for effective processing of the ore are all examples for a possible uneconomical mineral deposit. Most of the mineral deposits in the world are usually classified as **Geopotential**, which includes occurrences and deposits that have not yet been identified today, but can be expected to be discovered by exploration in the future. Comparable geological structures can already provide an indication of which areas might be encouraging for new discoveries of mineral deposits, and thus these areas can also be termed as "favorable" (Fig. 2.3).

The distinction between reserves, resources and geopotential is gradational and, in the end, whether a deposit can be economically exploited or not is dependent on the market prices for the relevant commodities. The resources and geopotential of today can be transferred into reserves by price increases, improvements in technology or increased exploration. The resources, which previously could not be economically exploited, can then be economically mined. In contrast, a decrease in prices or increase in costs, for example because of enhanced environmental requirements for mining or increased taxes, can result in reserves being downgraded into resources. The Kropfmühl graphite mine in Bavaria clearly exemplifies the dynamic of this interaction: the mine was closed in year 2000 because of economic considerations, and the reserves were downgraded to resources. The mining operations were restarted in 2012 as the economic conditions improved[2]—and resources were again upgraded to reserves.

New reserves are continuously being discovered, even today. In such cases, the geopotential is changed to a reserve. This is exemplified by the development of new offshore marine natural gas or oil fields that have increased significantly until now. The proportion of offshore oil production in the total global oil production has increased from 5% in 1950 to 40% today. In particular, the technological advances in drilling and production have contributed to this increase in offshore production. Today it is possible to drill in water depths of up to 3000 m—for example in the Gulf of Mexico. In 1950 production was possible at a maximum water depth of only 20 m.

---

[2]A general increase in the graphite price and an increased demand for its application in lithium-ion batteries contributed to an improvement in the economic conditions (Regiowiki [3]).

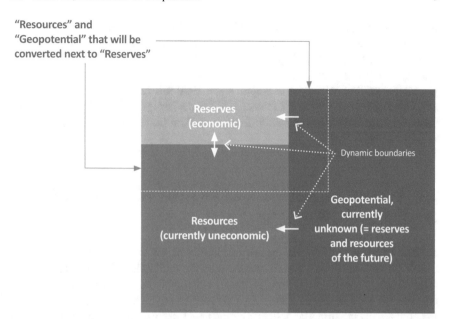

Fig. 2.3 Resource Box. The arrows indicate that new reserves and resources are discovered by exploration, and that the classification of mineral deposits as either resource or as reserve can change according to the economic conditions, such as fluctuating commodity prices (Modified from Scholz et al. [27])

Many of the occurrences classified as geopotential in 1950 are today either reserves or resources.

Even though the consumption of many raw materials continues to increase, resources and geopotential are always being transformed into resources because of economic and technical developments, and therefore there continues to be a sufficiency of natural resources. The reserves increase with consumption, and sometimes increase at an even greater rate than the global consumption. This effect was not previously taken into consideration—even by the Club of Rome, which is an international non-governmental organization founded in 1968 by leading industrialists, engineers, economists and scholars of the natural sciences and humanities. The Club of Rome suggested in its study "The Limits to Growth" that scarcity of natural mineral resources and environmental degradation would lead to serious crises and, before 2100, to a collapse and return to simple living conditions [4]. The effect of reserves growing with consumption was also not taken into account in the recent update of this study by the futurologist Randers [5].

The oil industry provides a further example of the concurrent increase of reserves: in 1950 oil production was about 540 million tonnes per year, and the known reserves at that time of 11.3 billion tonnes were sufficient for only about twenty years at the same rate of production. By 2013 the production had increased to 4.2 billion tonnes per year and the reserves amounted to 219 billion tonnes. Although the production in 2013 was more than seven-times greater than 1950, the reserves have increased disproportionately more and, based on the assumption of a constant production of 4.2 billion tonnes per year, they are now sufficient for more than 50 years. The same observation of increasing reserves can be made from the ratio of reserves to annual production of the metals: for example, since 1950 this ratio for zinc has remained at a value of between 20 and 25 years, and for copper it is about 40 years although the copper production has increased from about 4 million tonnes in 1960 to about 18 million tonnes in 2013.

2.3   Critical Raw Materials and Raw Materials of Strategic Economic Importance

Specialists often identify **critical** raw materials and raw materials of **strategic economic** importance during their discussions on the availability of raw materials. The term "critical" does not refer to the physical or chemical characteristics of the relevant element, or to the size of the resources and reserves, but on the availability of the raw material and its importance for the economy [6, 7]. For example, some raw materials are only mined in a few countries such as China, Chile or the Democratic Republic of Congo. The supply of these raw materials could be hindered by political crises or trade barriers, so that they are scarcely available on the global market. Although the geological reserves would be sufficient, the supply of the raw materials can be restricted. Each situation must be treated on a case-by-case basis because there is no standardized and objective "criticality threshold", beyond which a raw material can be regarded as critical.

A criticality matrix is usually used to evaluate the criticality[3] of raw material (Fig. 2.4) [8, 9]. If the availability of specific raw material poses a relatively high threat to the economy, then they are normally classified as "critical". This is the case for those raw materials that cannot be sufficiently recovered by recycling or substituted by other raw materials, and that are primarily sourced from foreign countries. The most comprehensively researched material flow analysis, which was undertaken at Yale University, USA [10, 11], introduced the concept of a third axis for environmental implications.[4] The European Union (EU) also applies an Environmental-Country-Risk factor in their risk analyses. However, the environmental risk factor

---

[3]The term "criticality" with reference to raw material supply has become established in the English-speaking world. It should be noted that in this context the use of the term "criticality" has nothing in common with the classical definitions in physics.

[4]The definition of this evaluation criterion is provided in Graedel et al. [10, p. 1066]. Manufacturers, government authorities and non-governmental organizations should demonstrate with this criterion the environmental impacts caused by the extraction processes of a specific raw material. The environmental impacts take into consideration the potential damage for the environment and mankind. The database is derived from the ecoinvent Centre—Swiss Centre for Life Cycle Inventories [12]. With the use of the third axis, a criticality space replaces the criticality matrix. See Graedel et al. [10, 11] for further details.

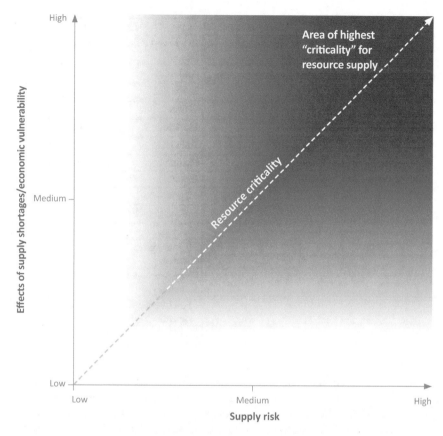

Fig. 2.4 Criticality Matrix for Raw Materials [8, 9, 15]. The criticality of a raw material results from the interplay between the supply risk, or the reliability of the supply-side, and the economic importance, of demand-side dependence on the relevant raw material. The gradational coloration indicates the fluid transition between critical and non-critical

developed by the team from Yale University evaluates of how close countries are to established policy goals. The EU applies this Environmental-Country-Risk factor [8] to the mining countries and use it as a measure of the degree to which the environmental performance in the producer countries could endanger the supply to the EU of specific raw materials—and is therefore not an analysis of the environmental risks, but of the supply risks.

The raw materials that are of special economic and political importance to the EU have been classified in two studies for the European Commission as particularly important for industry. These studies were undertaken in 2010 and 2014 and list 14, respectively 20, critical raw materials, and they are known as the "EU-14-Criticals" [8] and, respectively, the "EU-20-Criticals" [9] (see Table 2.1). China is the principal source for a large proportion of these raw materials (see Sect. 4.3.1, Box IV).

Table 2.1 Comparison of the EU-14-critical raw materials from 2010 and with the EU-20-Critical raw materials from 2014 [8, 9]

Critical raw materials in the EU-14 and EU-20 lists		
Only the EU-14 list	EU-14 and EU-20 lists	Only the EU-20 list
Tantalum	Antimony	Borate
	Beryllium	Chrome
	Cobalt	Coking coal
	Fluorite	Magnesite
	Gallium	Rock phosphate
	Germanium	Silicon metal
	Indium	
	Magnesium	
	Natural graphite	
	Niobium	
	Platinum group elements	
	Heavy Rare-earth elements	
	Light rare-earth elements	
	Tungsten	

The comparison of both lists demonstrates that the studies only present a snapshot of the raw material-supply situation, and this is continuously influenced by current events [13] and can change at any time. For example, the 2010 list only identifies 14 raw materials, and one of them (tantalum) was not included in the 2014 list, but seven other raw materials have been added. Chromium is included in the 2014 list for the first time, although in the 1970's it was practically always at the top end of criticality studies [14] but then rarely appeared in the studies undertaken in the first decade of the 21st century. (In 2017, EU list was updated: Seven mineral, one gaseous, and one agricultural raw material were added: barite, bismuth, hafnium, helium, natural rubber, elementary phosphorous, scandium, tantalum and vanadium. Two mineral raw materials were deleted from the list: chrome and magnesite. Coking coal is a borderline case. Although it narrowly misses the economic importance threshold, the EU kept it on the list for 2017 for sake of caution) [15].

The example of tantalum very clearly demonstrates the variability of the supply risk: the supply risk for a raw material is demonstrated by a matrix with which the supply situation for this raw material can be analyzed. The so-called country risk, with which each of the producer countries are classified, is shown on the Y-axis. The country risk is derived from the World Governance Index of the World Bank (WGI).[5] The following rule is generally applicable: the lower the WGI, the greater the country risk. The country risk is compared to the supply concentration

---

[5]See glossary World Governance Index of the World Bank (WGI).

according to the Herfindahl-Hirschmann-Index (HHI)[6] that is shown on the X-axis. The risk classification for tantalum during the period from 1996 to 2009 is shown as an example in Fig. 2.5. The smaller the market, then the short-term changes are proportionately greater, and this can significantly affect the studies on criticality. For many years, the most important source countries for tantalum were Australia, Brazil and Canada with high WG-indices and various African countries with negative WG-indices. The fluctuations in the proportion of production from these countries and/or changes to the risk-classification of a country can result in the variations in the evaluation of the supply risk.

Although these evaluations and lists of critical raw materials only provide a short-term view of the situation, they are an important guide for correctly estimating the necessity for action to secure the raw materials supplies for the future. The EU list of critical raw materials was therefore an important basis for the selection of eligible raw materials within the framework of the exploration funding program coordinated by the federal government [16]. The Federal Ministry of Education and Research (BMBF) in its recent "r⁴" research program *Raw Materials of Strategic Economic Importance for High-Tech Made in Germany* defined the group of "raw materials of strategic economic importance" as those raw materials (see also Sect. 3.3.2) that must be reliably available for the technologies of the future. These raw materials are required for the manufacturing of leading-edge products, and even small quantities have a major leverage effect on the economy and its added value [17, p. 51]: steel alloy metals, refractory raw materials, raw materials for the electronic industry and other high-technology raw materials such as rare-earth elements or the platinum group metals.[7]

A recent study by the German Mineral Resources Agency (DERA) investigates the effects of the supply risks of various intermediary products in the higher levels of the value-added chain, and not just on the risks from primary resources of raw materials [18]. The intermediary products are those from the processing or refining stages over and above the first phases of exploitation such as ore, concentrate, refined or pure metal, and they may be traded. This is important because German businesses in the metal and industrial minerals sectors increasingly use intermediary products, rather than primary mineral raw materials, as the basic material for their manufacturing processes. Until now there is scarcely any information on the supply risks of these globally traded goods that is available to these companies. Criticality analyses, such as that from DERA, should be continued up to the higher levels of the value-added chain.

---

[6] See glossary Herfindahl-Hirschmann-Index (HHI).

[7] A group of raw materials are termed "of strategic economic importance" with respect to an open technology concept (no restrictions because of the availability of raw materials) in an attempt to avoid the disadvantages of a static list (refer to BMBF [17]).

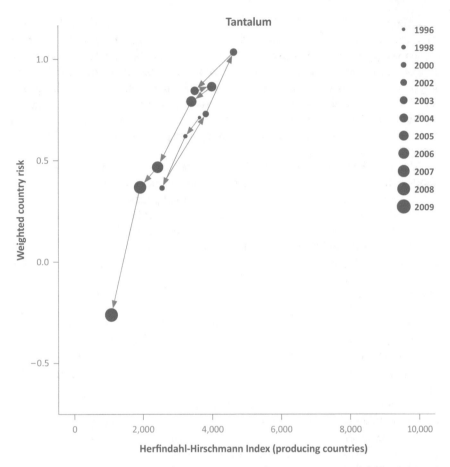

Fig. 2.5 Changes of the critically in the supply for tantalum during the period from 1996 to 2009 (With the friendly assistance from L. A. Tercero Espinoza. Modified from Sievers et al. [28], p. 8). As a result of the changes in the relative proportion of production from different source countries, which are weighted according to the WGI, the supply risk in the past is characterized by major fluctuations, and most recently has fallen significantly

2.4   Availability of Raw Materials: The Feedback Control Cycle of Raw Materials Supply and Commodity Studies

In order to ensure that there are sufficient raw materials available for the energy systems of the future, the future trends of their availability must be investigated now. In addition to the necessity for the occurrence of suitable mineral deposits, it must be recognized that exploitation, processing and use of mineral resources always has an impact on the environment and social structures. The effects on climate or the impairment of habitats, for example, are increasingly important in the evaluation of the availability of mineral raw materials. With respect to the future supply of raw materials, the overriding objective for mineral exploitation will be to minimize so far as is possible the impacts on the environment. At the same time, the quantities of exploitable geogenic[8] mineral resources are limited (compare Sect. 2.2). This is particularly the case for the fossil energy resources as these are actually consumed because the energy that is released by their combustion can only be used once,[9] whereas both the non-metallic and, in particular, the metallic raw materials can be recycled.

The availability of the energy natural resources is also determined by the Energy Returned on Energy Invested (EROI) factor, which describes the efficiency of the use of the energy sources and is defined as the ratio of the recovered energy to the energy invested.[10] This is determined from the ratio of the combustion energy of the exploited fossil fuel to the quantity of energy that is required for its exploitation. Obviously, if the exploitation of the fossil fuel consumes almost as much energy as is released by its combustion, then the exploitation is not economic. The value of the EROI factor varies from deposit to deposit.

The availability of renewable energy natural resources also depends on how much plant-based biomass can be grown. Furthermore, the availability of biomass as a source of energy is constrained because biomass also serves other purposes—for example for the nutrition of people and animals (see Sect. 4.3).

The consumption and availability of mineral raw materials are different from that of the fossil fuels and renewable energy resources. Mineral raw materials are exploited by mining (primary extraction) and are then processed in numerous ways for industrial purposes . They are, in other words, transferred from the geosphere to

---

[8]Geogenic means that the mineral raw materials are developed and occur naturally on Earth. This includes the raw materials groups of the mineral natural resources and fossil energy natural resources as they are defined here. They are distinguished in this context from the biologically grown resources such as biomass.

[9]Although the energy received is preserved and, for example, transferred to heat disseminated to the environment, the thermodyamic possibility for useful work is lost. Specialists describe the capability for useful work as exergy.

[10]Invested energy usually includes the total life cycle including the development and decommissioning of the necessary facilities (cumulative energy requirement).

the technosphere. Mineral raw materials can, in principle, be recovered by recycling back from the technosphere (secondary extraction).[11] The metals therefore present a more or less inexhaustible resource potential. Since metals in the technological environment are used, but not consumed, with optimization of the recycling rates they could almost be considered as renewable. At the rate at which the global quantities of metals processed by industry are increasing, the recycling of metals from secondary materials in the technological environment will also increase (see Sects. 3.4.4 and 3.4.5). In summary, the availability of mineral raw materials must therefore be evaluated with respect to both the geosphere and the technosphere.

2.4.1   The Feedback Control Cycle of Raw Material Supply

Although there is today a fundamental sufficiency of raw materials to meet the demand, specific market conditions can result in shortages and significant increases in commodity prices such as, for example, the widespread implementation of new technologies, periods of stronger economic growth (economic cycles) that are associated with an increased demand for raw materials, or financial speculation. In general, these price increases are offset by incipient market mechanisms in both the supply and demand sides. The systematics of the raw materials' commodity markets are shown in Fig. 2.6 that illustrates the feedback control cycle of raw material supply as defined by the Federal Institute for Geosciences and Natural Resources (BGR) [19]. Both the supply- and demand-side generally react to raw material shortages and price increases so that on the supply-side, for example, more natural resources can be recovered by improved efficiencies in recycling technologies, and on the demand-side the shortages are compensated by a more economical use of the relevant natural resources [20]. The Club of Rome study [4] and its current update do not sufficiently assess the principle of the feedback control cycle [5].

2.4.2   Studies on Raw Materials Criticality

Since the rapid increase of natural resources consumption in China at the beginning of the 21st century, and the similar increases in the commodity prices on the global markets, the issue of natural resources has returned to the political agenda. During the past ten years, various national and international advisory committees have undertaken numerous studies on the supply of potentially critical raw materials, and the future availability of these raw materials is analyzed and extrapolated—also for those raw materials that are considered to be important for the energy systems of

---

[11] Recycling is sometimes constrained by chemical changes of some raw materials, as for example lime that is burned into cement. These raw materials are generally sufficiently available, and are not critical to the energy transition.

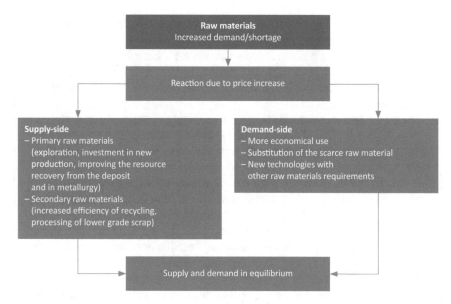

Fig. 2.6   The Feedback Control Cycle for Raw Material Supply (Modified from Wellmer and Dalheimer [29])

the future. In order to compare the various studies with each other it is necessary to consider in detail, on the one hand, how each study determines criticality and, on the other hand, how the reserves will be developed in the future.

BOX I: By-Product Elements

According to the feedback control cycle for raw material supply, raw material shortages and price increases will counteract each other because the market (supply and demand) regulates itself with time. This control cycle for raw material supply is only partially valid for those raw materials that do not form ore deposits on their own—these are the so-called by-product metals. By-product metals occur in deposits of other principal mineral raw materials (primary raw materials). The ore minerals of the primary ore (i.e. metal) and minerals containing the by-product metal can occur together as intergrowths, and in some cases the by-product element is contained in the crystal lattice of the primary ore mineral. The primary and by-product metals can often only be separated by using a significant input of energy. The extraction of a by-product metal is therefore inevitably coupled with the extraction of the primary metals, although this necessity only applies for the first stages of exploitation, the mining and processing,[12] of the so-called primary exploitation. The possibility of separating the by-product metal into truly marketable materials during the subsequent smelting process is dependent on the economics of the additional processing required. Many of the high-technology metals such as, for example, indium, germanium, gallium, tellurium or selenium, are only exploitable as by-products.

That the feedback control for raw material supply is only in part applicable for the by-product metals has the following reason: if the price for a by-product metal increases, there is generally little incentive for the mining company to increase its production at the relevant mines because the production will always depend on the demand for the primary metal that is produced. At best the smelting company, which metallurgically recovers the metals from the ore, may react to the price increase and increase the recovery of the by-product metals. For example, germanium is recovered as a by-product from zinc mining but an increase in the price of germanium is unlikely to lead to increase in the mine production of zinc. This economic relationship in mining is known as a "price inelasticity for by-product metals". At best the zinc smelter may be incentivized to recover more germanium from the zinc concentrates. One advantage of this situation is that the lead times for increasing the production of by-products are significantly less than for the primary products. For example, it may be necessary to undertake additional exploration for new deposits of primary raw materials in order to increase the resource base, but it is relatively simple to increase the production of the by-products just by optimizing the exploitation of the primary raw material. The facilities for recovering the by-products are often at the same location as the smelter for the primary raw material, and this can also be viewed as an additional advantage because it is not necessary to construct a new facility for by-product recovery at a greenfield site, and therefore avoid the often lengthy permitting procedures.

In the meantime, it is clear that the supply risk of raw materials is much more volatile if the market is small. The constraints of the studies on criticality are also very variable. As an example, two important and commonly cited studies have been compared [21]—namely the criticality study of the US American National Research Council (NRC) [22] and the study on the EU-14 Critical Raw Materials [8]. The results of this comparison demonstrate that the important conclusions are consistent (Fig. 2.7). However, there is a significant difference of opinion for manganese: although both studies rate the economic importance of manganese as very high, the supply risk is estimated as high by the NRC study but as low by the EU study. This probably reflects a different political evaluation of the country risk of one of the major producing countries—namely Gabon, where the manganese mining is controlled by a French mining company.

---

[12]Processing is undertaken in facilities located at the mine where an enriched and marketable product is produced by crushing, milling and various enrichment methods.

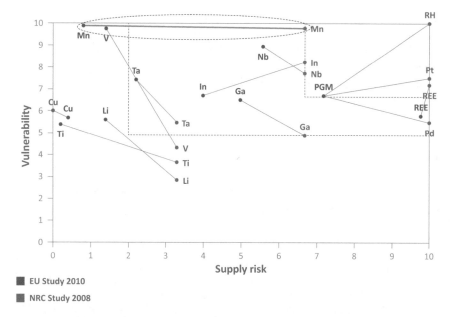

EU Study 2010

NRC Study 2008

Fig. 2.7 Comparison of the criticality evaluations of the US American National Research Council (NRC) in [22] with the EU-14 in 2010 (Modified after Erdmann and Graedel [21, p. 7626]). Significant differences in the estimation of the supply risk, for example for manganese (Mn), might be the result from dissimilar evaluations of the reliability of important producer countries. (Modified after Erdmann and Graedel [21], with permission of the American Chemical Society)

It is surprising that raw materials studies are often criticized because some of the raw materials that are classified as critical have never demonstrated supply problems in the past, and platinum is cited as an example [13]. The 20th century has even been described as the "mass grave of the forecasts" [23]. These criticisms, however, fail to recognize that the objective of these analyses must be to provide alerts for impending supply bottlenecks. The economy requires reliable information to be able to avoid, in advance, potential supply difficulties—for example by stockpiling sufficient raw materials or by preparing alternative technological strategies. If these alerts are taken seriously by the general economy, then it might be possible to prevent the predicted supply problems from even occurring. The forecast has in this case fulfilled its role—even if the predicted event did not actually occur because of the preventive measures that were taken. Over-optimistic forecasts that ignore the problems would be fatal. It is therefore desirable that a more pessimistic forecast is ultimately demonstrated to be incorrect or even falsified [24], rather than the forecast becomes reality and results in real supply problems. Many of the forecasts from the 20th century proved to be incorrect, but on the one hand this could have been because of political scenarios that did not actually materialize, or on the other hand because of the industry reacted effectively to the possible scenarios. Today the rare-earth elements are classified as critical in all the studies, but this is now the case because the industry did not give sufficient importance to the earlier forecasts [25]. Despite these earlier forecasts, the industry continued to rely on supplies from China, even though similar situations (creation of a near monopoly by means of dumping-prices so that many competitors were marginalized, and subsequent increase of prices by imposition of export restrictions) had been observed earlier, for example the cases of tungsten and fluorite (refer Sect. 3.4.2, Fig. 3.17).

References

(Note: All Web links listed were active as of the access date but may no longer be available.)

1. Bradshaw, A. M./Hamacher, T.: "Nuclear fusion and the helium supply problem". In: *Fusion Engineering and Design*, 88, 2013, S. 2694–2697.
2. Hagelüken, C.: "Secondary Raw Material Sources for Precious and Special Metals". In: Sinding-Larsen, R./Wellmer, F.-W. (Eds.): *Non-Renewable Resource Issues- Geoscientific and Societal Challenges*. Dordrecht, Heidelberg: Springer Verlag, 2012, pp. 195–212.
3. Regiowiki: *Graphit Kropfmühl GmbH*, Regiowiki für Niederbayern und Altötting 2015. URL: http://regiowiki.pnp.de/index.php/Graphit_Kropfm%C3%BChl_GmbH [accessed: 28.09.2015].
4. Meadows, D. H./Meadows, D. L./Randers, J./Behrens III, W. W.: *The Limits to Growth*, A Report for the Club of Rome's Project on the Predicament of Mankind, Universe Books: New York 1972.
5. Randers, J.: *2052 – A global Forecast for the next forty Years*, A Report to the Club of Rome commemorating the 40th Anniversary of the "The Limits to Growth", White River Junction, Vermont: Chelsea Green Publishing 2012.
6. Buchholz, P./Huy, D./Sievers, H.: "DERA Rohstoffliste 2012, Angebotskonzentration bei Metallen und Industriemineralen – Potenzielle Preis- und Lieferrisiken", In: *DERA Rohstoffinformationen Nr. 10*, Deutsche Rohstoffagentur in der Bundesanstalt für Geowissenschaften und Rohstoffe 2012.
7. Bradshaw, A. M./Reuter, B./Hamacher, T.: "The Potential Scarcity of Rare Elements for the Energiewende". In: *GREEN*, 3: 2, 2013, S. 93–111.
8. European Commission: *Critical raw materials for the EU* (Report of the Ad-hoc-Working Group on defining critical Raw Materials), Brussels 2010. URL: http://ec.europa.eu/enterprise/policies/raw-materials/files/docs/report-b_en.pdf [accessed: 01.05.2014].
9. European Commission: *Critical raw materials for the EU*, Report of the Ad-hoc-Working Group on defining critical Raw Materials, Brussels 2014. URL: http://ec.europa.eu/enterprise/policies/raw-materials/files/docs/crm-report-on-critical-raw-materials_en.pdf [accessed: 26.06.2014].
10. Graedel, T. E./Barr, R./Chandler, C./Chase, T./Choi, J./Christoffersen, L./Friedlander, E./Henly, C./Jun, C./Nassar, N. T./Schechner, D./Warren, S./Yang, M./Zhu, C.: "Methodology of Metal Criticality Determination". In: *Environmental Science & Technology*, 46, 2012, p. 1063–1070.
11. Graedel, T.E./Harper, E.M./Nasser, N.T./Reck, B.K.: "On the Materials Basis of modern Society". In: *Proceedings of the National Academy of Sciences of the USA* (Early Edition), 2013. URL: http://www.pnas.org/cgi/doi/10.1073/pnas.1312752110 [accessed: 01.04.2014].
12. Hischier, R./Weidema, B. (Eds.): *Implementation of Life Cycle Impact Assessment Methods* (Data v.2.2, ecoinvent report No. 3), Dübendorf: Swiss Centre for Life Cycle Inventories 2010.
13. Buijs, B./Sievers, H./Tercero Espinoza, L. A.: "Limits to the critical raw material approach". In: *Waste and Resource Management*, 165: WR4, 2012, pp. 201–208.

14. Wellmer, F.-W./Schmidt, H.: "Versorgungslage bei Rohstoffen". In: *Stahl und Eisen*, 89: 2, 1989, pp. 55–60.
15. European Commission: Communication from the Commission to the European Parliament, the Council, the European Economic and Social Committee and the Committee on the Regions on the 2017 list of Critical Raw Materials for the EU, 13.9.2017. URL: https://ec.europa.eu/transparency/regdoc/rep/1/2017/EN/COM-2017-490-F1-EN-MAI N-PART-1.PDF [accessed 26.1.2018].
16. Bundesministerium für Wirtschaft und Energie: *Bekanntmachung im Rahmen der Rohstoffstrategie der Bundesregierung: Richtlinien über die Gewährung von bedingt rückzahlbaren Zuwendungen zur Verbesserung der Versorgung der Bundesrepublik Deutschland mit kritischen Rohstoffen (Explorationsförderrichtlinien)*, Berlin: BMWi 2012.
17. Bundesministerium für Bildung und Forschung: *Wirtschaftsstrategische Rohstoffe für den Hightech-Standort Deutschland (r^4)*, Bonn: BMBF 2012.
18. Buchholz, P./Huy, D./Liedtke, M./Schmidt, M.: *DERA-Rohstoffliste 2014, Angebotskonzentration bei mineralischen Rohstoffen und Zwischenprodukten – Potenzielle Preis- und Lieferrisiken* (DERA Rohstoffinformationen Nr. 24), Berlin, Deutsche Rohstoffagentur in der Bundesanstalt für Geowissenschaften und Rohstoffe 2015. URL: http://www.deutsche-rohstoffagentur.de/D ERA/DE/Publikationen/Schriftenreihe/schriftenreihe_node.html [Stand: accessed 04.2015].
19. Wellmer, F.-W./Becker-Platen, J.D. (Eds.): *Mit der Erde leben – Beiträge Geologischer Dienste zur Dasesinsvorsorge und nachhaltigen Entwicklung*, Heidelberg: Springer Verlag 1999.
20. Wellmer, F.-W./Hagelüken, C.: "The Feedback Control Cycle of Mineral Supply, Increase of Raw Material Efficiency, and Sustainable Development". In: *Minerals*, 5, 2015, pp. 815–836.
21. Erdmann, L./Graedel, T.E.: "Criticality of non-fuel minerals: a review of major approaches and analyses". In: *Environmental Science and Technology*, Bd. 45: 18, 2011, pp. 7620–7630.
22. National Research Council of the National Academies: *Minerals, Critical Minerals, and the U.S. Economy*, Washington, D.C.: The National Academies Press 2008.
23. Sames, C.-W./Kegel, K.-E./Johannes, D.: "Millennium – 100 Jahre Bergbau im Rückblick". In: *Erzmetall*, 53: 12, 2000, pp. 759–760.
24. Honolka, H.: *Die Eigendynamik sozialwissenschaftlicher Aussagen: Zur Theorie des self-fulfilling prophecy*, Frankfurt a. M.: Campus Verlag 1976.
25. Bundesanstalt für Geowissenschaften und Rohstoffe/Deutsches Institut für Wirtschaftsforschung: *Auswirkungen der weltweiten Konzentrierung in der Bergbauproduktion auf die Rohstoffversorgung der deutschen Wirtschaft* (Joint Report for the Research Project Nr. 26/97, Gutachten im Auftrag des Bundesministeriums für Wirtschaft), Bundesanstalt für Geowissenschaften und Rohstoffe, Deutsches Institut für Wirtschaftsforschung 1999, pp. 253–269.
26. Achzet, B./Reller, A./Zepf, V./Rennie, C./Ashfield, M./Simmons, J.: *Materials critical to the Energy Industry. An introduction* (Report for the BP Energy Sustainability Challenge), Augsburg: University Augsburg 2011. URL: http://www.physik.uni-augsburg.de/lehrstuehle/rst/do wnloads/Materials_Handbook_Rev_2012.pdf [accessed: 28.10.2014].
27. Scholz, R./Wellmer, F.-W./DeYoung jr., J. H.: "Phosphorus Losses in Production Processes before the 'Crude Ore' and 'Marketable Production' Entries in Reported Statistics". In: Scholz, R./Roy, A.H./Brand, F.S./Hellums, D.T./Ulrich, A.E. (Eds.): *Sustainable Phosphorus Management – A Global Transdisciplinary Roadmap*, Dordrecht, Heidelberg etc.: Springer Verlag 2014.
28. Sievers, H./Buijs, B./Tercero Espinoza, L. A.: "Critical Minerals for the EU", In: Polinares Working Paper 31, 2012. URL: http://www.polinares.eu/docs/d2-1/polinares_wp2_chapter19. pdf [accessed: 15.10.2015].
29. Wellmer, F.-W./Dalheimer, M.: "The Feedback Control Cycle as Regulator of past and future Mineral Supply". In: *Mineralium Deposita*, 47: 7, 2012, pp. 713–729.

Chapter 3
Supply of Raw Materials and Effects of the Global Economy

The availability of raw materials is influenced by the supply- and demand-sides. Furthermore, the supply of raw materials—in general and including those for the new energy systems of the future—is dependent on developments in the global mining business and global economy. Political guidance can play an important role in this situation.

3.1 Primary Exploitation of Raw Materials and Forecasts of Future Availability

After they have been mined, mineral raw materials are also distinguished by how and under what economic circumstances they can be extracted from the ore. As a result, the following groups are identified:

1. Raw materials that (a) are extracted from ores that occur in deposits of that specific raw material, such as iron, copper and gold, and (b) are extracted as by-products, such as indium or germanium (from zinc deposits), tellurium (from copper deposits) or rhenium (from copper deposits, together with by-product molybdenum), and
2. Raw materials that are mined by (a) major mining companies and (b) mid-size or small mining companies.

Major mining companies are becoming increasingly focused on the so-called Tier-One projects, which are the large and long-lived projects with low production costs and high cash flow.[1] Typical raw materials include coal, iron ore, cop-

---

[1] Cash flow within a business, and is the surplus of receipts into over disbursements from the company.

per, nickel, zinc, gold, diamonds, potash and phosphate [1]. The turnover derived from exploitation of the refractory metals such as tantalum and tungsten or the rare-earth elements is generally too low to be of interest to the major mining companies. The mid-size and small mining companies tend to control the smaller niche markets.

The consequences of this situation for exploration include the following: major mining companies develop new mineral deposits at the same rate at which they exploit reserves. Exploration is therefore a continuous process, even though it is driven by economic cycles and/or exploration success. Systematic exploration such as this for a specific raw material is undertaken by the big companies that are specialized in just one or two commodities, such as for example uranium, potash or phosphate companies. The mid- and small-size mining companies are also interested in maintaining their reserve base, but they tend not to commit to long-term systematic exploration. Instead, they wait for opportunities to acquire indications of new deposits[2] of raw materials that have been discovered by small exploration companies. These small exploration companies include the so-called junior companies that do not have a cash flow, but finance their exploration activities by issuing their stock in countries such as Canada, USA or Australia—that is exclusively by loaned capital and not from sales of their own products. They tend to focus on raw materials that are currently "fashionable", including those that are characterized by high prices (such as the rare-earth elements during recent years) or have a favorable image among stock investors, such as for example diamonds. The exploration is therefore strongly dependent on favorable commodity prices or "fashion" issues, and the exploration cycles are therefore short-term. The feedback control cycle of raw material supply is therefore particularly relevant due to the interplay of activities on both the supply and demand sides (see Sect. 2.4).

The exploration activities of all companies, with the exception of the junior companies, in the sector are financed from their profits. The profits are very dependent on the variations of the metal prices, and the price trend is again coupled to the cycles of economic development and therefore also behaves cyclically. The exploration expenditures are thus closely correlated to the price trends, and generally follow the metal price cycles with a phase delay of one or two years (Fig. 3.1). Furthermore, it must be realized that, like many other areas of scientific research, the understanding about new deposits does not increase continuously. Usually discoveries are made at irregular intervals, and in exploration they can be compared to a scientific breakthrough. The costs also increase with increasingly more precise methods for the exploration of deposits. This means that the early phases of exploration are relatively inexpen-

---

[2]Indications can include areas ranging from prospecting of very favorable geological formations (areas that, based on experience possess a high potential for discovery of deposits), through outcrops with favorable ore grades, to explored occurrences or even mineral deposits.

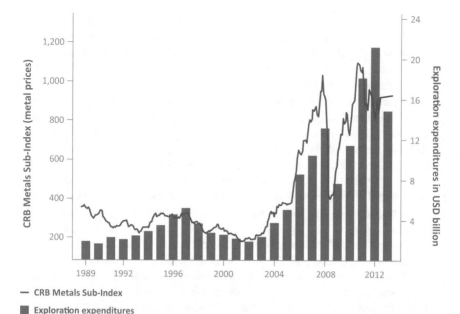

Fig. 3.1 Development of metal prices and exploration expenditures in the period from 1989 to 2013 (from BGR [2]: data from CRB [3] and MEG [4]). The extreme swings of the economic boom (2003–2007 and after 2008) and bust (2008) phases dominate the gradual course of the commodity cycles. They also demonstrate the commodity hype at the beginning of this century

sive, whereas the precise definition of the quality of a deposit in the later phases, for example with numerous drill holes and possibly even underground exploration activities, is more cost intensive. This cyclical effect must be taken into consideration in order to understand why there are always periods with a relative lack of new discoveries, and why supply problems for certain raw materials are forecast in some studies.

BOX II: The Ratio of Reserves to Consumption as an Early-Warning Indicator for the Availability of Raw Materials

The ratio of reserves to consumption, which is incorrectly used as a guide to the measurement of the future availability of raw materials and known as the "static range" (this is the ratio of reserves to mine production,[3] Fig. 3.2 and Appendix A, Fig. A.2), does not represent the absolute "life-expectancy" of a raw material but is only a snapshot of a dynamic system. All the parameters, such as reserves, resources, consumption and production, are continuously changing so that current figures cannot be the basis for projections into the future of developments in the sector. At best, the reserve/consumption ratio may be suitable as an early-warning indicator [7]. If this value decreases to less than about 10–15 years, it could be regarded as a critical value because of the pre-production development time (see Sect. 4.1.1, BOX XIII). The monitoring of the long-term movement of the ratios is the best method to recognize warning signals about the long-term supply of raw materials. This reflects the fact that there is no institution in the world with the capability of evaluating all the mineral deposits on Earth. Published reserves are always only the sum of individual deposits that have been defined by mining companies and, in some cases, state authorities.

Many of the high-technology raw materials that are important to the development of future energy systems do not occur in their own discrete deposits. For raw materials that occur as by-products in deposits with other natural resources, the ratio of reserves to consumption is only valid to a limited extent as an early-warning indicator. With respect to the availability of these by-product raw materials this ratio cannot provide any indication of future availability, since only that proportion of the primary raw material, from which the by-product is extracted, can be used as the basis for the calculation. In general, this reflects the demand for the by-product raw material and ignores the potential for additional extractable resources, and therefore this leads to an underestimate of the availability.

The calculated ratio of reserves/consumption is often too low for these raw materials because the underlying database is generally poor. The reason for this is that the necessary data is either not systematically collected by the mining companies, and/or not published. The data required include ore grade and recovery grade of the by-product raw material in the relevant primary resource. On the other hand, the potential of previously mined deposits from which the by-products were not extracted, must be evaluated. An exemplary statistical analysis on the potential of germanium has recently been completed. It demonstrates that the reserve/consumption ratio of the primary raw material provides a possible way for obtaining a more reliable estimate of the actual availability of the by-product raw materials. The incomplete data about the grades of the by-product elements in the ore and the recovered grades, for example the proportion of germanium in the global smelter output from zinc concentrates that is actually recovered, suggests that in general this proportion is currently much less than 100% for all of the high-technology raw materials [8].

---

[3]From the long-term perspective, consumption of primary raw materials and mine production are very similar. Differences occur because of stockpiling.

Reserves represent the working capital of the mining companies, which require an appropriately current database for their project planning or, in other words, to determine when, at what production rate, and at what cost can a mineral deposit be exploited. The normal time horizon is usually short- to medium-term (up to about 50 years), so that deposits are of no interest if they can expect to be exploited in one hundred years of even later.

The "static ranges" for tin and antimony are estimated to be seven and eleven years respectively. These values are close to being critical, bearing in mind that the pre-production development time for new mining projects is currently 10–15 years.[4] Antimony is included in the EU-20 list of critical raw materials because of the high supply concentration from one source country (China), the restricted possibilities for substitution, and the limited possibilities for recycling. However, neither tin or antimony are relevant for the energy systems of the future.

The "static range" (ratio of reserves to consumption) can be regarded as a measure of necessary research and development. Zinc and copper require continuous exploration in order to guarantee sufficient ore for exploitation that will meet the demand requirements and maintain a balance between reserves and production. For raw materials with a high value for the "static range", for example phosphate or potash, the requirements for additional exploration is relatively low. This is particularly the case for phosphate, which has a "static range" of almost 300 years [9, p. 118ff] and sufficient reserves are known. The "static range" ratio, as already noted, does not provide any indication of the availability of the raw material but can be related to the types and tonnages of the deposits that typically host the various raw materials. Low ratios, such as for copper and zinc, are typical for raw materials with lens-like, individually separate, ore bodies. Raw materials, such as bauxite (aluminum raw material, see Appendix A, Fig. A.2), phosphate or coal, typically occur in beds or seams so that sample sites can be widely extrapolated, and have moderate long-term ratios with values often exceeding one hundred years. The ratios for iron and bauxite have dropped during recent years, which is a direct result of the rapid increase in Chinese consumption since 2002. As already explained, exploration cannot keep pace with rapid increases in consumer demand (global iron ore production in 2000 was 1010 million tonnes, and in 2013 it as 3110 million tonnes; global bauxite production in 2000 was 127 million tonnes, and in 2013 it was 283 million tonnes) [9, pp. 26f and 84f; 10, pp. 28f and 82f].

---

[4]The time series of this early-warning indicator for the most important raw materials, derived from the period 1988–2012, are provided in Appendix A, Fig. A.2.

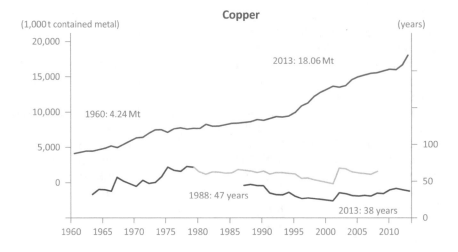

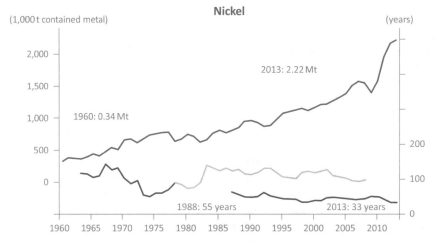

— Mine production (BGR)

— Static range based on reserve base (USGS)

— Static range based on measured and indicated reserves (USGS)

**Fig. 3.2 The "static future availability", or ranges, of copper and nickel based on the relevant
reserves** (The reserves are based on the information published by the US Geological Survey (USGS)
in its annual reports, the Mineral Commodity Summaries. The reserves are classified according to
their level of reliability. "Measured" means that the reserves are calculated from a high density of
sample data relative to the extent of the deposits and their grades, and can therefore be regarded as
relatively reliable. "Indicated" are reserves calculated from a less dense sampling grid as compared
to that for "measured" reserves, but the quality of the data is sufficient for interpolation of the deposit
characteristics between the sample points, but because of the lower number of sample points the
level of reliability is lower.) **compared to the actual mine production** [5, 6]. The decreasing trend
of the ranges with contemporaneously increasing production rates is caused by the strong increase
in demand since the beginning of the 21st century. The increase in demand is primarily related to
the economic upturn of China. Exploration is not able to replace the mined reserves at a sufficiently
fast rate

3.1.1 Production Peak Instead of Static Life Time of Reserves and Resources?

The production peak, or the time of maximum production of a raw material, is another attempt to assess the availability of raw materials. The production with time of a raw material from a defined mining district usually follows a characteristic curve. After the initial discovery of the deposit, the production increases exponentially until the point of maximum production, the so-called production peak, is reached, and then decreases (Fig. 3.3). The annual production history of specific raw material within a defined area, be it a mining district, a country, a continent or the world, always follows a similar curve. The curve varies for each natural resource and each deposit according to the mining methods and the developments in the markets, and only very occasionally can it be mathematically defined and symmetrical. The curve is named after the US American geologist, Marion Kind Hubbert, who used it in 1956 to predict that the oil production peak in the USA (excluding Alaska) would be reach in 1971 [11]. At that time, Hubbert's predictions were almost correct, and in fact was only one year in error. However, from the current perspective, Hubbert had miscalculated because he did not take into his consideration the so-called unconventional oil deposits[5] in the USA, which are now being developed with the newest technologies. If the unconventional deposits are included in oil production statistics, then the production peak of oil (or Peak Oil) has not yet been reached (Fig. 3.4). This is the case not just for the USA, but throughout the world the oil production maximum is shifting further into the future by taking the unconventional deposits into account.

The discussion about global peaks has been advanced for other raw materials, in particular for phosphate that is an essential plant nutrient, cannot be substituted, and is not available in virtually unlimited quantities such as nitrogen from the air or potash from seawater. Unfortunately, the calculation of a peak often mingles two different effects: a supply-driven peak, which is the most relevant for a discussion on the availability of raw materials, and a demand-driven peak. The 1970 oil production peak in the USA, excluding Alaska, is an example of a supply-driven peak. The supply-driven peak demonstrates the total amount of a natural resource that can be made available. The demand-driven peaks cannot provide any reliable indication of the actual physical availability of a raw material because they are always variable for different reasons, for example because of environmental regulations that might prohibit the use of a specific raw material. The asbestos peak is an example

---

[5]The term "unconventional" does not pertain to the raw material, but to the type of deposit in which the natural resource occurs. The term "unconventional deposits" of the hydrocarbons includes deposits of oil and natural gas where the raw material is not hosted by permeable rocks, as is the case for conventional deposits. For unconventional deposits, special drilling technologies (horizontal drilling) and hydraulic stimulation to fracture the reservoir rocks around the drill hole are generally required to ensure that the natural gas or oil can freely flow towards the drill hole and be recovered. Occurrences hosted by shales or coal seams, the so-called source rock, nowadays belong to this deposit type. Tight-gas deposits, or those that are hosted by impermeable sandstones or carbonates, are now regarded as conventional because of the experience obtained about this deposit type since the 1990s.

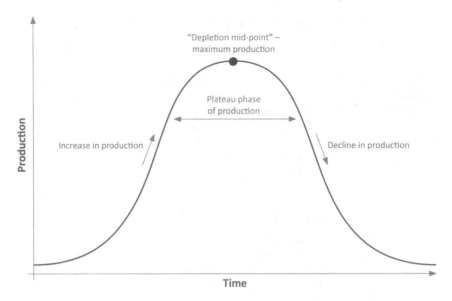

Fig. 3.3 The bell-shaped "Hubbert" curve (schematically after Hubbert [11]). This reflects the ideal curve of the production of a non-renewable natural resource in a defined area or deposit. The area under the curve reflects the volumes of the available natural resource. In the ideal case, as shown here, the production peak correlates with the 50% level of exploitation of the deposit, or the so-called Depletion Mid-Point (dmp)

for this—although today asbestos is available (supply side) there is practically no demand because of its carcinogenic effects which virtually prohibit its application. The demand-driven peak therefore does not provide any information about the actual availability of the raw material. In addition to environmental regulations, rationalization can also result in a reduction of the demand for a natural resource.

Fundamentally it is only possible to predict a peak for a specific natural resource if the total geological potential (Fig. 2.3) of that natural resource is, at least statistically, understood. The global total quantity of a natural resource is known as the Ultimate Recoverable Resource (URR), but until now it has not been possible to estimate a URR for any raw material. The reason for this is that global exploration has not yet covered the whole surface of the earth, let alone exploration below 300 m depth that has only been carried out in locally defined areas. For a while, some experts claimed that it would be possible to estimate a URR for oil because the major sedimentary basins in the world that contain the oil deposits were known and well researched. However, today increasingly more unconventional oil deposits are being developed and this makes it significantly more difficult to estimate a URR.

Finally, there is little point in discussing "Peak Minerals" [13] because the quantity of reserves and therefore the timing of the production peak can always shift. If the demand or commodity prices increase, or if new and more efficient and therefore more cost-effective technologies are developed for the mining production processes,

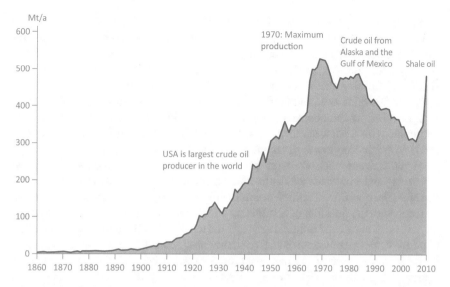

Fig. 3.4 The oil production curve for US America (in million tonnes per year) (modified and updated from Wellmer [12, p. 128]). The production peak predicted by Hubbert is clearly shown as being one year in error (1970), and the current developments (production of shale oil) that Hubbert had not taken into consideration because he was not aware of the today's technological advances

then the profit will also increase. As already discussed, this can result in an increase in exploration and the so-called cut-off grades[6] that mark the threshold of economic mineability will also shift. Resources and geopotential are upgraded to reserves and so, with price increases and new technological developments, the reserves generally increase disproportionately, and the peak time moves further into the future. This is clearly demonstrated by the example of the recent developments to extract oil from primary deposits by fracking (oil shale; see Sect. 3.4.3, BOX VIII). In contrast reserves can revert to resources if the prices decrease and the exploitation is no longer profitable—for example because the use of expensive drilling technologies makes the exploitation of unconventional deposits such as oil shales too costly.

3.1.2 Energy and Water Requirements

Economic and political events are not alone in influencing the availability of raw materials. It is also vitally important that the mines are supplied with sufficient

---

[6]Metal deposits are usually zoned according to the grade of the raw material, and this is reflected in the profitability of the exploitation. A boundary must be established to define the limits of ore that can be mined. This is the cut-off grade or threshold. The lowest cut-off grade is that at which the operating costs are just covered (for example, see Wellmer et al. [14]). There is an extensive literature on the optimization of the cut-off grade.

energy and (fresh) water, which is required for the processing of the ores, and this can be a significant challenge. Many mining camps are located in desert areas, for example the copper mines in northern Chile, which is the largest copper producer in the world. Some experts [15] consider the availability of water to be the most significant limiting factor for the future supply of natural resources. Brackish or salty water is often used in the mines located in arid or semi-arid regions. There is a major effort by the copper mining industry in Chile, for example, to reduce the requirement for fresh water in ore processing. Of the planned projects in Chile during 2011, 46% intended to use desalinated water and 23% salt water Rosas et al. [16].

Significant amounts of energy are required for the desalination of water, and this energy must be made available. Salt water is found practically everywhere, and therefore the question of sufficient availability of water for mineral exploitation can be reduced to the question of a sufficient and sustainable supply of energy. Energy, together with social and political issues, is the most critical factor for the exploitation of primary mineral raw materials. Even if the energy costs for mining should decrease in the future because of the increasing or even exclusive use of renewable energy sources, there will always be an appreciable total cost for the supply of energy from renewable sources. The reason for this is because the construction of the facilities for producing renewable energy requires significant amount of relatively expensive raw materials, such as metals. The specific costs for materials required for renewable energy facilities are currently higher than those required for conventional facilities [17–19]. As shown in Sect. 3.2, the real (inflation corrected) prices for nearly all commodities have remained more or less the same during the past hundred years. In the long-term a total internalization of all costs, including for example the costs related environmental impacts, must be expected to result in increased prices, even in real terms. Additionally, the energy used for exploiting the raw materials used in the facilities will also increase. It must be expected that the metal grades in the mines will continue to decrease, that the deposits will be deeper, and the ores more complex and difficult to process.

Metal grades are still a long way from the mineralogical limit at which metals are incorporated in crystal lattices of the rock-forming minerals (silicates) and not as metal sulfides, from which they are easier to extract, and therefore can only be recovered with much higher energy input; the mineralogical limit is therefore also an energy threshold [20]. Despite this, the concept of an energy balance, or an "energy amortization", must today be taken into consideration.

The amount of energy used in the exploitation of mineral raw materials is also important with respect to climate change and the energy transition. For example, the question of the CO_2 balance of green technologies such as wind-powered generators or solar cells must be addressed. This also depends on how much energy must be consumed to exploit the raw materials that are required for the construction of the energy-generating facilities. This could effectively reduce the savings obtained by the generation of "green electricity" (see Sect. 3.4.3).

The energy balance of mining companies can in the future be improved if they adapt their production to the changeable availability of electric power derived from sun and wind. It is conceivable that those companies, which are dependent on renewable energy sources, will adapt their production to the more cost-effective power generated during surplus off-peak periods. There is currently no market for this surplus energy, and it could be used to process very low-grade ore that requires a more intense technological input. Today many open-pit mines have two stockpiles: a high-grade and a low-grade ore stockpile. The latter is processed during periods of favorable high commodity prices.

The energy balance can be calculated for fossil energy sources, as is well demonstrated in the following example. The cumulative energy balance(VDI [21]: replaces the 1997 version) for the production from the Gifhorn Trough oil and gas field in Germany was calculated according to the VDI (Association of German Engineers) Guideline 4600, and this demonstrated that only 0.8% of the energy content in the oil production and only 0.4% of the energy content in the natural gas production was used for the development, production and supply of energy from the field [22]. This represents an "Energy Returned on Energy Invested", or "EROI", factor of about 100 for oil and nearly 200 for natural gas. The production from the Gifhorn Trough therefore requires a relatively low input of energy. This type of balance calculation depends on numerous parameters of the hydrocarbon field, such as the depth to the production zone or the flow rate of oil or natural gas to the bore hole, and these can be much less favorable in other oil and gas fields. For example, the EROI values in North America for crude oil are 20–40, and for natural gas are 15–25 [23].

Some of the ideas about the production of biomass raise the issue of energy balance at the other end of the energy efficiency spectrum (see Sect. 4.3.6). For example, the irrigation of desert areas with desalinated sea water in order to grow biomass for energy production is almost a "zero-sum game", because the total energy required for the desalination is about fifty percent of that produced from the biomass.

Primary Exploitation of Raw Materials Resources and Forecasts of Availability

Possible supply risks for some raw materials are not due to geological availability but to insufficient availability in the market. The demand-driven increases in commodity prices generally occur during economic upturns. These result, with a delay, in increased investment in the mining sector that in the long-term increases the supply and matches the increased demand. However, the increase of supply resulting from new discoveries occurs very slowly: the development of a new mining project, from the discovery of the deposit to the commencement of production, requires lead times of an average of approximately ten years. Lead times are also required to increase the capacity of onsite production facilities. As a result, this causes short-term price and delivery risks in the supply of raw materials.

Exploration activity generally depends on the profits and commodity price expectations of the mining companies. Exploration activity is therefore subject to economic development cycles. Exploration success is also cyclical and is not necessarily linked to the level of exploration expenditures, and exploration efficiency must therefore be evaluated over the long-term.

The ratio of reserves to consumption does not reflect, as is so often assumed, the expected life of the reserves, but is always only just a static snapshot of a dynamic system. The ratio is also dependent on the type and size of the mineral deposits. It can be used as an indicator of the exploration expenditure that is necessary. It can also be a warning signal if the ratio approaches a value of ten, which is the typical lead-time for a new mining project. This is currently the case for antimony and tin, which are two raw materials that are relatively unimportant for the energy systems of the future.

Raw materials peaks are often of general interest to the public, in particular the question if and when the geological availability of a raw material has reached its maximum, and is then followed by shortages. The geopotential of almost all mineral raw materials is unknown, and therefore raw materials peaks cannot be predicted. Furthermore, it is technically possible for the mineral raw materials, particularly the metals, to be recycled. These raw materials are therefore not consumed, but only used. The peaks are related to changes in the demand, and therefore the discussion on Peak Minerals is generally not very purposeful.

The exploitation of mineral natural resources is dependent on the availability of water and energy. Salt water, which is accessible virtually everywhere, can be desalinated with a certain expenditure of energy. The issue of water supply is therefore reduced mainly to a question of energy generation. With respect to the raw material requirements for the "green technologies", a sustainable approach to the water and energy supply for mineral exploitation is a central challenge for the future. A transition to renewable energy technologies is therefore also necessary for mineral exploitation.

3.2   Price Setting and Market Mechanisms

In contrast to manufactured products, mineral raw materials do not have a unique selling point. Although there are different qualities of oil, basically a mineral raw material is a single bulk product. The commodities are traded globally based on physical and chemical specifications, usually in commodities future exchanges such as base metals on the London Metal Exchange (LME) or crude oil on the International Petroleum Exchange. As a result, small shortfalls or covering somewhere in the market can lead to significant price variations. Because the economic cycles are always accompanied by fluctuations in the demand for raw materials, the markets are only rarely in equilibrium, and most fluctuate between the buyer and seller markets.[7]

---

[7] A buyer's market is a market where the buyer determines the price because there is an oversupply and the buyer can put pressure to decrease the prices; and a seller's market is just the opposite.

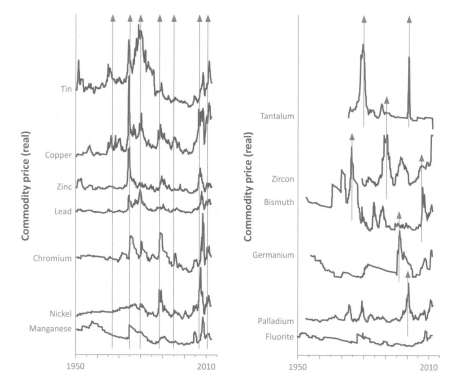

Fig. 3.5 Price peaks of "normal" commodities and specialty metals (Buchholz [24], based on Bräuniger et al. [25]). The prices are shown as real prices (inflation-corrected); the price peaks for the "normal" base metals and steel alloy metals (left) generally occur at the same time, whereas the curves for the specialty metals display individual price peaks. The global economic development trends, such as the price boom from 2003 or the finance crisis in 2008, can be recognized in this comparison

These market instabilities are emphasized by high production capacities that can only react slowly to changes in demand. The economic cycles equally affect all the normal commodities such as the base metals or steel alloy metals, and it is therefore no surprise that price peaks of the "normal" mineral commodities occur at approximately the same time (Fig. 3.5, left) [25].

In contrast, the price behavior of the specialty metals, such as tantalum, bismuth, germanium or palladium, is determined by events specific to the individual raw materials. This includes, for example, changes in the demand related to new technological processes or new products in the market that require specific individual metals for their manufacture. Therefore, no time-equivalence can be recognized in the price behavior of these metals (Fig. 3.5, right). Sudden changes in price occur relatively often in this small sector of the market.

The trends in the market can also be dominated by the demand of single market players. For example, the recent relatively prolonged increase in demand from China

Fig. 3.6 Development of nominal and real metal prices since the Second World War, based on the CRB Metals sub-index [26]. The general declining trend of the real commodity prices (blue arrow) is occasionally interrupted by economic and political events that resulted in sharp increases in prices. Over the long-term the real prices have remained more or less at the same price level since the Second World War

was a factor that determined the market trends. Previous peak price cycles generally occurred at a frequency of about four to six years. The price boom that ended in 2013 had started in 2003, and was briefly interrupted by the global financial crisis in 2008. This unusually long boom in raw materials will therefore be part of a new commodity super-cycle.

The nominal metal prices, in particular those of the metals traded on exchanges, increase over the long-term. However, the real prices, or the prices corrected for inflation[8], decrease with time (Fig. 3.6). Economic and political events repeatedly interrupt this decreasing trend with upward spikes. The real prices increased very strongly during, for example, the oil crises in the 1970s or more recently during the raw materials boom that was initiated in China in 2003. Over the long-term, from 1950 to today, the real prices have remained at an approximately constant level (Fig. 3.6).

The development of the commodity prices over the past hundred years is surprising. The commodity prices increased sharply with the beginning of the First World War because of the enormous requirements of the armaments industry. At the end of

---

[8]The nominal and the real prices of commodities must be distinguished. Nominal prices are those that are quoted every day on the market (often exchanges). The real prices are adjusted for inflation, using the relevant inflation indices, from a selected year, the so-called basis year. At the time of the basis year the real and nominal prices are the same. The evaluation of real prices is important to identify trends in a time series.

the war, industrial production decreased and the demand was suddenly reduced. This was also followed by various economic crises and a major recession in the mining industry, and the prices of the metallic raw materials decreased. This situation and a rationalization of the mining industry with the introduction of modern technical processes, such as for example the use of more powerful open-pit equipment (excavators, trucks), resulted in a total restructuring of mining over the following years. As a result, a price level was reached that has not significantly changed until today. Although many people have a different perception, the real prices of nearly all commodities have hardly increased since the end of the First World War. During this time, there were some price peaks caused by inflation, and this is well demonstrated by price development of the non-ferrous metals (Figs. 3.6 and 3.7). There are no indications to suggest that this situation will change for most raw materials, at least not in the medium-term. A constant price level is, however, conditional upon a continuing increase in the productivity of mining by new technical developments. This increase in productivity must not be to the detriment of the continuing compliance with environmental and social standards, which are both important fundamentals to ensure the social acceptance for mining and mineral processing over the long-term (see Sect. 3.4.2). Ultimately the complete internalization of all costs related to environmental impacts must have an upward effect on commodity prices [29].

The fluctuations of the commodity prices are currently increasing in frequency, the so-called price volatility that is measured by the standard deviations of the monthly or annual fluctuations. Many experts consider that this effect is related to speculation in the stock markets and metal exchanges, in particular in the Exchange Traded Funds (ETF).[9] The traders on the commodity exchange markets normally buy and sell metals forwardly, without actually physically owning the metals. The market volumes are therefore often twenty- to sixty-times the volumes of physically traded metals. With time, the volatility also displays cyclic trends.

These price movements are essential to a market economy, and they provide the incentives to the industry that ensures a reliable supply of raw materials. The feedback control cycle for raw material supply was described in Sect. 2.4 and plays an important role in the markets [30]. Price peaks that frequently occur for the specialty metals, such as the electronic metals, are often the trigger for these events. The production cycles are becoming increasing shorter and new products, which may require different raw materials, are released quicker to the market, and this results in short-term germanium fromsupply shortages. In these circumstances a mindset ("hype") that spreads quickly through the markets can play a significant role. For example, the opinion or mindset can prevail that a critical supply shortage of a raw material is to be expected. Afterwards it is often clear that sufficient quantities of this raw material were always

---

[9]Exchange Traded Funds are investment funds that are traded on the stock markets. There are both passively managed and actively managed ETFs. Passively managed ETFs are based on an index, like the German market index DAX. Actively managed ETFs invest on the basis of the managerial experience in stocks or also commodities, so that they actually buy, hold and sell commodities. These ETFs do not only buy, hold and sell contracts for the future delivery of the metals, but also metals themselves (for example with the title to physical metal). They therefore act on the commodity markets as both the customer as well as the supplier.

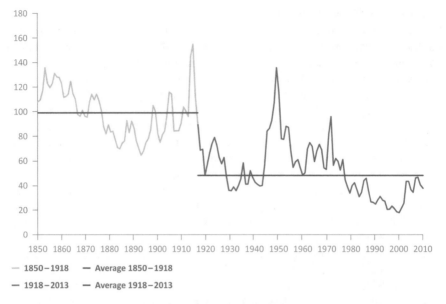

— 1850–1918   — Average 1850–1918
— 1918–2013   — Average 1918–2013

Fig. 3.7   Index of real prices of non-ferrous metals from 1850 to 2013 (normalized to 1900). The prices for aluminum, copper, lead, zinc and tin are weighted by means of the real value of the production. The data is derived from the London Metal Exchange and its predecessors (after Stürmer [27] and Stürmer [28]: the corrections for inflation are based on the British consumer price index). The offset after the First World War is the result of wartime production, economic crises, and rationalization in the mining industry (*source* Stürmer [27] and [28], with permission of the author)

available. In the final analysis, every company should be prepared for fluctuating commodity prices and fluctuating volatility. The industry should develop long-term strategies to ensure that they will have a reliable supply of raw materials at any time, and that they are not following short-term opinions ("hype") that influence the markets.

Price Setting and Market Mechanisms

As a general rule, commodities are globally traded and are qualitatively similar bulk products, regardless of their source. The differences in quality of untreated or unrefined products, such as crude oil, are compensated by discounts and premiums on the price. The price is therefore set in a global market.

Although the nominal commodity prices are increasing, over the long-term the real prices (corrected for inflation) are remaining around the same level. This long-term trend can continue in the future if the challenge of the expected increase in demand for raw materials can be countered by technological improvements and enhanced productivity. In the short-term, however, minor shortfalls or surpluses can result in significant price fluctuations. For some specific raw materials, the long-term price trend can be interrupted by sudden increases in demand, for example coupled with the development of new products. Commodities that only have a small market are characterized by relatively erratic price fluctuations because only a few individuals can dominate the market activity. Speculation and mindset perceptions (hypes) can also affect the price trends and are in part responsible for the recent significant increase in price volatility.

3.3   The Demand-Side

Today's commodity market is global, and there are scarcely any local commodity markets remaining. There are a few exceptions that may arise, for example, by restrictions on exports. The natural gas supplied through pipelines is also a special case since it is, in a certain way, restricted to specific regions. However, in this case globalization is also developing as natural gas is being increasingly transported by ship throughout the world in the form of liquefied natural gas (LNG).

Today almost any means of wealth creation requires several different raw materials that are sourced from different parts of the world, and this contributes to globalization. Even the service sector can no longer operate without raw materials, mainly because of the computers and telecommunications hardware such as smart phones, for which electronic components and microchips are essential. Computer chips today, for example, consist of up to sixty elements [31]. Some semi-conductor elements such as gallium are indispensable for smart phones. There was scarcely any demand for these elements 25 years ago, but today they are essential for the communication technology in telephones. These examples demonstrate how the demand for raw materials is primarily determined by advances in technology, and this demand changes with the development of new products and technologies that are associated with social progress and industrialization. Technologies of the future are important drivers of this trend, and many products such as automobiles or smart phones are in demand throughout the world and are produced globally. As a result, almost every state or national economy is served from the same global pool of raw materials. Studies that evaluate the availability of raw materials for the energy systems of the future must also consider the development of the global economy as well as that of the technologies of the future.

3.3.1   General Trends

The development of new products during the past decades demonstrates that requirements of raw materials for national economies depends on their industrialization and economic development. In this context, industrialization is a process that transfers the factors of production[10] from primary production (mining, agriculture and forestry, fisheries) into the industrial sector (processing). This is associated with an increase in material input as well as the material intensity (the ratio of the raw material input to the Gross Domestic Product, GDP, that is also known as Intensity of Use Factor) [32].

From a certain stage of economic development, the proportion of industrial production in the GDP decreases and the tertiary sector (service sector), which has a relatively low material intensity, increases in importance. Although the absolute input of raw materials increases, the material intensity decreases. The level of industrial development of a country is therefore reflected in the demand for raw materials as compared to its economic productivity. If the material intensity is compared to the GDP per person, then this results in a typical bell-shaped curve that is presented exemplarily in Fig. 3.8 based on copper in several industrial countries.

With the increasing industrialization of the countries in the Organization for Economic Co-operation and Development (OECD), the raw material demands have risen strongly since the 1950s. This trend, as already seen for China, is also observed in many of the emergent developing and transitional countries. As a result, more raw materials have been used in the past fifty years than in the previous history of mankind. However, the structure of global raw material supply has fundamentally changed during the first decade of this century. Almost up to the end of the previous millennium about 25% of the world's population, those from the industrialized countries, required about 70–80% of the global raw material production (except for coal). At this time, the consumption of raw materials by the People's Republic of China increased very suddenly (see Fig. 3.9b). In 1990 the Chinese proportion of global steel consumption was only 8%, and by 2012 this had increased to almost 46%. The demand for other metals also increased very sharply: the proportion of copper consumption rose from 6 to 43%, and aluminum from 5% to almost 44%. China is now the largest consumer in the world of all the main raw materials, apart from oil and natural gas for which it is in second place of the USA.

The curves, which demonstrate the increase of metal consumption (Fig. 3.8), are learning curves. With time countries "learn" to manufacture increasingly more valuable products from the same quantity of raw materials. Initially the raw materials requirements increase very strongly during the industrialization phase because the development of infrastructure and construction of production facilities for the manufacturing industries is very material intensive. As the infrastructure becomes better developed, the requirement for raw materials decrease so that from a certain moment in time the curve flattens out. The flattening of the curve commences with the peak of the material intensity bell-curve that is generally attained once the countries enter the phase with a high per capita income. There are other general rules. For example, during the development of a country it requires various raw materials successively after each other, and the sequence of reaching the maxima of the material intensities of the most important metals has been remarkably constant for the major industrial-

---

[10]Factors of production pertain to all material and immaterial resources and services that are required for the production of goods.

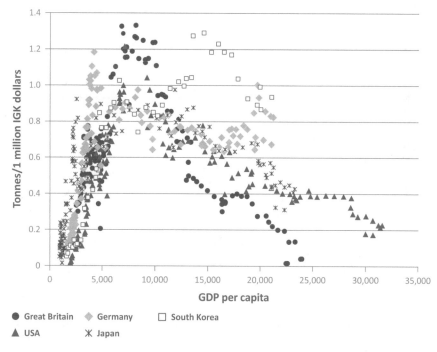

Fig. 3.8 Development of the material intensity of copper in selected industrialized countries [33, p. 26], (methodology after Kravis et al. [34]) based on the International Geary-Khamis dollar (IGK-$). (The IGK-$ is a comparative currency based on the USD calculated by the World Bank in order to be able to make comparisons between countries with different currencies. One IGK-$ represents the purchasing power of one USD at a specific point in time. 1990 is the base year for this graph. Other currencies are converted into IGK-$ according to their purchasing power. In contrast to the gross domestic product based on exchange rates, this method permits an appropriate presentation of the relative prices of goods and services). (*source* Stürmer and von Hagen [33], with permission of the authors)

ized countries: tin, zinc, crude steel, copper and aluminum. The average level of the material intensity is highest for crude steel, followed by aluminum, copper, tin and zinc [33]. If the material intensity then falls again then the consumption of raw materials stagnates but, in some cases, it increases further despite measures to increase the productivity of raw materials.

These universal rules provide guidance for general trend forecasts of the future availability of raw materials for the energy systems of the future:

Despite the sharp increase in the consumption of raw materials in China (Fig. 3.9a), the consumption curve for some raw materials can be seen to be flattening out (Fig. 3.9b). However, it is difficult to predict when the consumption will stabilize similar to that in the EU or the USA (Fig. 3.8). A major demand for raw materials is not expected from the other developing countries before 2020 [33]. However, after that the other more populated emerging nations and developing countries, such as

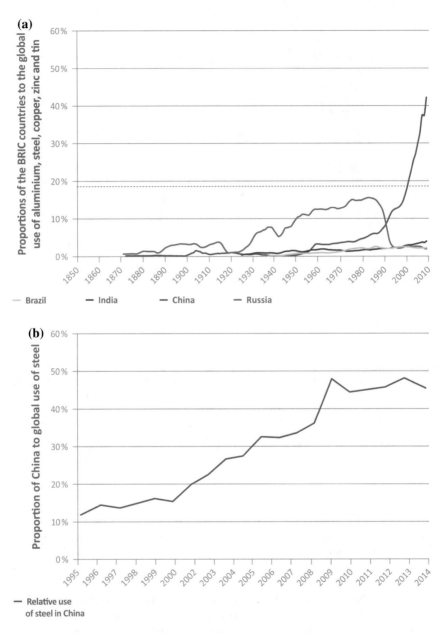

Fig. 3.9 Raw material input as an indicator of the level of industrial development of a country:
(**a**) comparison of the raw material inputs of aluminum, steel, copper, zinc and tin in the BRIC
countries (Brazil, Russia, India and China) reflects the enormous economic growth in China. The
raw material input is shown as the average proportion of the global demand [33, p. 35]. **b** The use of
steel in China, relative to the global demand, shows a flattening trend in recent years. This indicates
that, with respect to its use of steel, China has already passed its maximum of the material intensity.
(*Source* Stürmer and von Hagen [33], with permission of the authors)

India, Indonesia or Brazil, will follow the same development path as China. The timing of the sudden increase in demand for raw materials (the tipping point) is also not easy to predict, although the process will not have been concluded before 2050. A continuous increase in the competition for raw materials by the industrializing emerging nations and the developing countries must be expected during the next decades, and this will be associated in the future with increasing pressure on demand and price increases, even if also cyclical, for the raw materials. Fundamentally, the developments in the various countries and the technological changes are very difficult to predict, and therefore the future requirements for raw materials are also difficult to predict. Scenarios for the prediction and identification of possible problems in the supply of raw materials are therefore useful in any case.

3.3.2 Technological Developments in the Demand-Side

Germany, together with the USA, China and Japan, is one of the top industrial and trading countries in the world. Industrial (manufacturing businesses, except from the construction business) wealth creation in Germany during 2012 contributed proportionately 25.8% to the GDP, which is one of the highest proportions of any of the EU 28 countries, and only Romania was higher with a proportion of 32.5%. The average for the EU countries is 19.1%, with Italy on 18.4%, Great Britain 14.5% and France 12.5%.[11] All value-adding activities, in particular industrial wealth creation, require raw materials, and therefore Germany is one of the major consumers of raw materials. Germany in 2013 was the third biggest consumer of copper and aluminum in the world, fourth biggest of nickel and tin, fifth biggest of lead, sixth biggest of zinc and seventh of steel [2]. Germany is the third biggest export nation in the world after China and the USA, and the German strength is in high-technology products that require an increasingly broad range of raw materials. The high-technology strategy [36, pp. 12–13; 37, pp. 18–21] of the Federal government is therefore aiming to increase the economical and efficient use of raw materials through research, new technologies and the dissemination of innovations.

The automobile demonstrates the way in which our products have become increasingly more complex. In 1950, a car was a relatively simple product manufactured from steel, copper, lead, zinc, aluminum, rubber and plastics, but today it is a high-technology product, or a computer-on-wheels, for which numerous different raw materials are required (Fig. 3.10). Today's vehicle can contain up to 150 microprocessors with up to 6000 semi-conductors [39]. The high proportion of electronics in the vehicle, and particularly the development of the computer chips, are mainly responsible for this diversity in raw materials. According to the Intel company, a chip manufactured in the 1980s contained 12 different elements, and in the 1990s it contained 16 different elements, but since the beginning of 2000 it increased to over 60 elements [40, p. 38]. Apart from the automobile and computer industries, the

---

[11]Statistisches Bundesamt [35] [Federal Statistical Office (Destatis)].

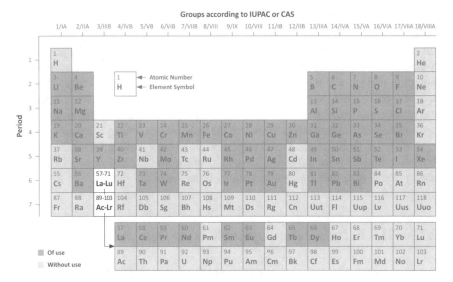

Fig. 3.10 The diversity **of elements used in automobiles** [38]

measurement and control technologies, which are vital for the efficiency of materials, are a further example for the consumption of an increasing number of raw materials. Fundamentally there are scarcely any elements that are only required for only one of the high-technology industries. The renewable energy technologies and other technologies that are required for the energy transition will therefore be competing with numerous other applications for these elements (Fig. 3.11).

Only rarely are the energy systems the most important application for individual elements. For example, permanent magnets constructed with rare-earth elements are used in wind-power turbines, but hard-disc drives in computers and other machines are the main application for these permanent magnets. The automobile, electronic and communication sectors are the main competitors for the raw materials that are required by the energy systems of the future [42].

Sudden and rapid changes in technologies can be reflected in the demand for raw materials as well as the prices. This is particularly the case in the consumer and entertainment sectors of the electronic business that is characterized by the most frequent changes of its products. The example of germanium (Fig. 3.12a) illustrates how strongly the applications and commodity prices can change. The commodity price movements are partly coupled to the changes in the technological applications (Fig. 3.12b), but also reflect general economic developments and political stimulus. The price increase in the late 1990s was primarily related to the increasing use of glass-fiber systems, including the production of solar cells. The price decline in 2009, however, was primarily due to the world financial crisis. There are several reasons, mostly political, for the most recent price increases since 2011. The supply-side was affected by the export taxes levied by China on germanium oxide, a production

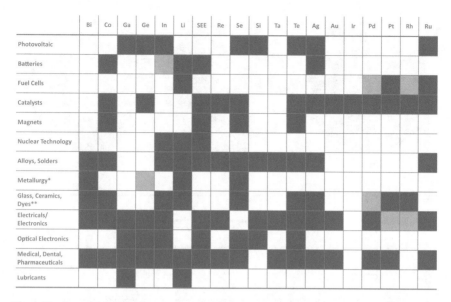

Fig. 3.11 Competition for raw materials between the energy technologies and other uses (modified from Hagelüken [41]). Dark-blue indicates that this is an important raw material for this application; light-blue indicates that the raw material is used for this application; and white indicates that there is no use for the raw materials. (*Additives in, for example, smelting, plating. ** Includes indium tin oxide (ITO) layers and coatings on glass)

facility in China was forced to close because of environmental concerns, and China announced that it would stockpile germanium for strategic reasons. The latter also probably resulted in increased speculation pertaining to germanium. At the same time, there was an increase in the demand related to, among others, the increasing use of germanium in light-emitting diodes (LED) and solar cells.

The television industry provides a good example of how quickly an industry can change. The transition from cathode ray tubes televisions to LCD televisions occurred within only two years from 2006 to 2007 [46]. The glass of a classical cathode-ray tube contained lead (in the throat and funnel sections) as a protection from X-ray radiation, and barium and strontium in the screen itself. After the introduction of today's LCD flat-screens, these elements were no longer required to produce the glass. The glass substrates of the modern displays consist of alumino-borosilicate glass. However, the technology transition to modern displays has resulted in a dependency on other critical raw materials: indium and tin. Both these raw materials are used in the transparent ITO (indium tin oxide) layers that function as electrodes in the displays.

The level of production during the exploitation of raw materials is generally, both for primary as well as for by-product raw materials, less flexible as compared to the changes in demand, and therefore there are rapid changes in the ratio between supply and demand. This is particularly the case for the specialty metals and for the raw materials used in electronic products, and this is the reason for the common price

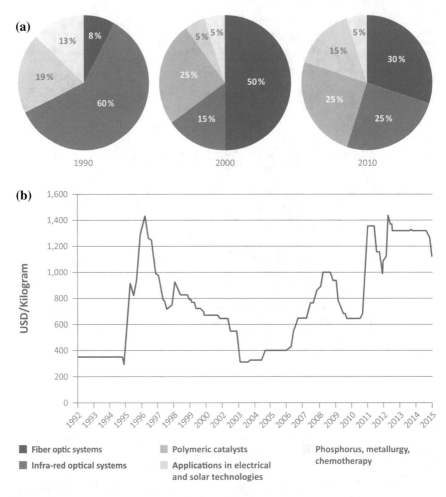

Fig. 3.12 Applications and price trend for germanium: **a** The example of germanium demonstrates that the demand for specific applications of a raw material can change radically because of technological developments, and this during a period in which the number of known germanium occurrences in the world has increased. The production of refined germanium in 1990 was 55 tonnes (tonnes of contained germanium), in 2000 it was 74 tonnes, in 2010 it was 120 tonnes, and in 2012 it was 128 tonnes [43]. **b** The price development of germanium from 1992 to 2015 (modified from Melcher/Buchholz [43]; updated after BGR [44, 45])

surges or even price peaks that is clearly illustrated by the case of germanium in Fig. 3.12b. These price peaks, however, are necessary to initiate the feedback control cycle of raw material supply, which is essential to a functioning raw materials industry in a free market economy.

Modern neodymium-iron-boron magnets in electric motors are an example of the raw materials required by an energy technology. The development of these strong permanent magnets permits the construction of electric motors that are more compact and less susceptible to failures. These robust electrical synchronous motors have become quickly established in numerous applications. For example, they are used as traction motors in vehicles or also for electrical kitchen appliances. China still has a near monopoly for the rare-earth elements that include neodymium. The supply situation has become strained by the increase in demand for neodymium or for the related dysprosium. The challenge to the industry is therefore to learn to be more flexible with the use of raw materials. Increased flexibility in the industry will be reflected by a reduction in its dependency on specific raw materials. It will then become less probable that price peaks occur, which could affect the availability of raw materials required for the energy systems of the future.

The extent to which industry can respond flexibly is demonstrated by an example from a steel works that protected itself from price movements of the alloy metals by appropriate changes to its applications (Figs. 3.13a, b). This is particularly important for energy systems since high quality steels of all types will be required for these systems.

An important basis for the systematic documentation of the demand for raw materials by the technologies of the future was prepared in 2009 for the Federal Ministry for Economic Affairs and Technology by the Fraunhofer Institute for Systems and Innovation Research (Fraunhofer ISI) together with the Institute for Futures Studies and Technology Assessment (IZT). Both institutes presented a study ([48, 49]; see also update: Marscheider-Weidemann et al. [50]). that analyzed in detail the supply situation and criticality for various raw materials. The German Mineral Resources Agency (DERA) plans to regularly update the demand trends.

Furthermore, research projects that focus on increasing the efficient usage and on the flexible use of raw materials are supported by the Federal Ministry of Education and Research (BMBF) program on "Innovations of Materials for Industry and Society" [51]. (WING), as part of the BMBF framework program "Research for Sustainable Development" [52]. (FONA), and as part of the federal government's high-technology strategy. These programs include the r^2 "Innovative Technologies for Efficiency in Raw Materials Intensive Production Processes" [53], r^3 "Innovative Technologies for Raw Materials Efficiency—Strategic Metals and Minerals", and r^4 "Raw Materials of Strategic Economic Importance for High-Tech Made in Germany" [54]. The following conclusion can be drawn from these studies about technology developments on the demand-side: developments on the demand-side will always, to a certain extent, be unpredictable. It is therefore necessary to develop wide-ranging solutions to ensure the availability of raw materials of strategic economic importance that are required for the energy systems of the future. Scenarios that evaluate technology developments and possible changes in the raw materials requirements can help in the understanding of the future.

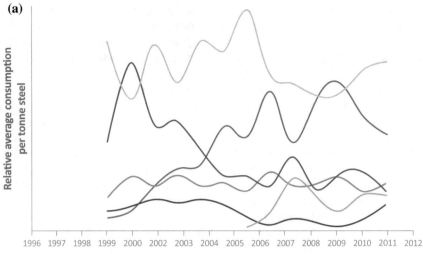

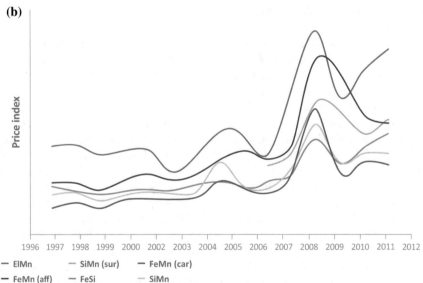

Fig. 3.13 Industrial flexibility towards price fluctuations—an example from the steel indus-try (modified from Lachmund [47]). **a** Use of various alloy metals. **b** Price movements of the alloy metals. The alloy additives are FeSi: ferrosilicon; FeMn(car): ferromanganese produced by carboth-ermal reduction (relatively high carbon content); FeMn(aff): low carbon ferromanganese; SiMn: silica manganese; SiMn(sur): very low carbon silica manganese; ElMn: electrolytic manganese (99% manganese, or practically pure manganese, with the lowest contents of carbon). (*Source* Lachmund [47], with permission of the author)

The Demand-Side

The raw materials resource industry is today almost completely globalized, and virtually every phase of social and industrial development is dependent on natural resources. The use of raw materials during the industrialization process is characteristically depicted by a bell-shaped curve.

The use of raw materials in the energy systems of the future is always in direct competition with the demand for raw materials by other sectors, which have arisen by the development of the global economy and from technologies of the future. These very rapid fluctuations in demand are often difficult to predict. It is therefore necessary for the industry to develop solutions to become flexible in its use of raw materials. Warning signals of supply risks can be identified in time by continuous monitoring of the supply and demand of raw materials, as well as through scenarios about the possible future development.

3.4 The Supply-Side

Today two general sources for mineral and metal raw materials are distinguished, on the one hand the primary raw materials derived from mineral deposits in the earth (the geosphere), and on the other hand raw materials recovered from recycling or other processes from industrial products (the technical environment, or so-called technosphere). It is practical to evaluate the raw materials supply separately for both categories, mainly because of the technical developments in the production processes of primary and secondary raw materials. In future, the requirements for mineral raw materials will continue to grow because of the increasing the global population and rising standards of living, the supply possibilities must be evaluated with respect to this growing demand.

3.4.1 The Impacts of Geology and Mining Economics on the Supply of Primary Raw Materials

Mineral deposits are always unevenly distributed within specific regions, and this must always be taken into consideration in the evaluation of production of primary raw materials. Some raw materials are therefore concentrated in only a few countries, and often only a few companies are involved in the exploitation of these raw materials. From the political aspect, this situation can lead to a danger of raw material supplies being interrupted by third-party actions, such as the export restrictions that were imposed by China for rare-earth elements in 2011 and 2012, and this can lead to distortions to the conditions for trading and competition in the markets [55, p. 21f]. There are five important issues that contribute to a concentration of supply:

1. **Regional Distribution of Deposits**: Each of the mineral and metal natural resource occurs in a typical geological setting. For example, crude oil that is formed from marine organisms only occurs in sedimentary rocks ("source rocks"), and not in magmatic rocks. Nickel deposits, in contrast, are associated with basic, magnesium-rich and silica-poor, magmatic rocks or their weathering products. The age of the rocks that host the mineral deposits is also an important criterion. For example, the richest iron ore deposits are associated with Banded Iron Formations that are practically all found in the oldest rock formations on Earth, the so-called shield areas on the continents such as are present in Australia, Brazil or Canada, but not in the younger rocks that are found in central Europe. Geology is unevenly distributed over the surface of the earth, and so the distribution of mineral deposits is also uneven. Furthermore, the number, size and quality of those deposits that do occur in regions with favorable geology are also irregularly distributed, and the reasons for this are still not fully understood (Fig. 3.14).

2. **Selection of Mineral Deposits**: The global mining industry operates in a global and open space without local markets, as is the case today, and the major mining companies naturally concentrate on the best mineral deposits, which leads to an additional uneven distribution of mining activity. As a result, not all the known mineral deposits are mined, but only the most lucrative. The so-called Lower Third Rule, by which the major mining companies generally try to only invest in new projects that, as compared to all other projects for similar raw materials, plot in the lower third of cost distribution. This can be seen as a protection against price fluctuations. The grades are unevenly distributed through most mineral deposits, and the overall economics can be maximized by commencing the mining in the higher-grade sectors of the deposits. The average grade then generally decreases with time. If this effect cannot be compensated by rationalization of the operations, then the less profitable mines are squeezed out of the market by new producers operating in the lower third, or at least the lower half, of the cost distribution curve. There is therefore a continuous economic selection process that generally leads to further regional concentration of the mining districts.

3. **Economic Concentration of Mining**: In the global economy, the acquisition and takeover of competitors is commonly observed, and this trend also occurs in the mining industry, where it has a special effect. The initial exploration for mineral deposits is a relatively inexpensive as it often depends on the initial "good idea" of where the most favorable areas can be found. Every company, newcomer or small-time prospector therefore has the same chances in the exploration for new deposits. New and innovative mining companies are always being founded, and it is estimated that these companies are responsible for the discovery of half of the new mineral deposits in the world. These companies are particularly innovative and flexible, and their work is usually more cost-effective as compared to the bigger companies, or majors. For example, all the diamond mines in Canada

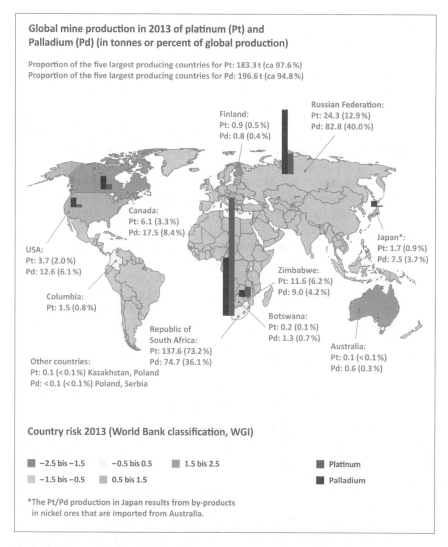

Global mine production in 2013 of platinum (Pt) and
Palladium (Pd) (in tonnes or percent of global production)

Proportion of the five largest producing countries for Pt: 183.3 t (ca 97.6 %)
Proportion of the five largest producing countries for Pd: 196.6 t (ca 94.8 %)

Finland:
Pt: 0.9 (0.5 %)
Pd: 0.8 (0.4 %)

Russian Federation:
Pt: 24.3 (12.9 %)
Pd: 82.8 (40.0 %)

Canada:
Pt: 6.1 (3.3 %)
Pd: 17.5 (8.4 %)

Japan*:
Pt: 1.7 (0.9 %)
Pd: 7.5 (3.7 %)

USA:
Pt: 3.7 (2.0 %)
Pd: 12.6 (6.1 %)

Zimbabwe:
Pt: 11.6 (6.2 %)
Pd: 9.0 (4.2 %)

Columbia:
Pt: 1.5 (0.8 %)

Botswana:
Pt: 0.2 (0.1 %)
Pd: 1.3 (0.7 %)

Republic of
South Africa:
Pt: 137.6 (73.2 %)
Pd: 74.7 (36.1 %)

Australia:
Pt: 0.1 (< 0.1 %)
Pd: 0.6 (0.3 %)

Other countries:
Pt: 0.1 (< 0.1 %) Kazakhstan, Poland
Pd: < 0.1 (< 0.1 %) Poland, Serbia

Country risk 2013 (World Bank classification, WGI)

| ■ −2.5 bis −1.5 | −0.5 bis 0.5 | ■ 1.5 bis 2.5 | ■ Platinum |
| ■ −1.5 bis −0.5 | ■ 0.5 bis 1.5 | | ■ Palladium |

*The Pt/Pd production in Japan results from by-products
in nickel ores that are imported from Australia.

Fig. 3.14 Reasons for the supply risk of raw materials, as illustrated by the examples of platinum and palladium (after Schmidt [56, p. 72ff]). The potential source countries are classified according to their country-risk based on the World Governance Index

were discovered by prospectors or small exploration companies. The same is true for the technology developments in the production of shale-hosted gas and oil that was also initiated by the small companies. However, these small pioneering companies are later taken-over by the majors, and the big mining companies often grow by acquisition of successful small companies. The Canadian diamond mines, for example, now belong to the two biggest Australian mining companies. Company acquisitions are especially successful in areas where the deposits are regionally concentrated. The companies that are already successfully active in the region are particularly suited for the acquisition of other smaller companies. These companies already have the necessary local infrastructure, and they can therefore develop new mines more quickly and efficiently. The regional concentration can therefore in the long-term transition into a concentration of the companies, or company.

4. *Cost Benefits*: Countries with a low-cost profile have significantly lower mining costs for most raw materials, and can successfully compete with, and ultimately squeeze out, competitors in other countries. For example, rare-earth element mining in China where the low level of the production costs is achieved through lower wages or reduced costs for environmental and social projects. The concentration of production also occurs in countries with approximately the same level of production costs and environmental and social standards. As already mentioned in Sect. 3.1, the major mining companies are increasingly focused on large, long-lived projects with low production costs and high cash flow, and these are known as the Tier-One projects [1]. If these companies are not willing to reduce their production during longer periods of low metal prices, and therefore stabilize the prices, their competitors with higher production costs will be squeezed out of the market. This situation occurred in 2015 in the Australian iron ore market.

5. *Concentration of the Smelters*: Not only a concentration of the mining activities, but also the concentration of smelters, can have an impact on the supply of raw materials for the energy systems of the future. This is especially the case for the by-product metals and the fabrication of intermediary products. German companies generally do not source raw materials of the first stage, but intermediary products that are used in the manufacture of high-technology items.

With respect to the importance of a reliable supply of mineral resources, it is essential to observe concentration trends that could endanger the free movement of raw materials throughout the world. It is therefore only consistent that the December 2013 [57]. Coalition Agreement of the Federal Government includes, under the heading "Expansion of Monitoring": "We will commission the German Mineral Resources Agency (DERA) to monitor the critical mineral raw materials, and regularly report on the availability of critical mineral resources for the German industry".

Looking towards 2050 and an ever-changing raw materials profile of the German national economy, there will always be situations whereby one country suddenly has a dominating position in the supply of a raw material or raw materials, and attempts to intervene in the international flow of trade and, despite all international trade agreements, to use mineral raw materials for its own political advantage. In future, it must be expected that the mining and smelting of raw materials will trend towards further concentration in the mining industry rather than diversification.

BOX IV: Concentrations and Country Risks

The concentration of mineral deposits, especially those of the rare natural resources, is determined by geology and, together with the trends towards a concentration of production, allow the formation of oligopolies and enable countries or companies to manipulate the commodity markets by, for example, export restrictions. This can result in deterioration in the reliability of supply of raw materials. This can be exemplified by two of the platinum-group elements, platinum and palladium: of the potential source countries, those with a medium to high country-risk such as South Africa or Russia often contain the best deposits and an especially high proportion of global production (Fig. 3.14).

In order to evaluate trends in concentration, the German Mineral Resources Agency (DERA) has developed a diagram (Fig. 3.15) that compares the regional concentration and company concentration using the so-called Herfindahl-Hirschmann Index (HHI) with the so-called weighted country-risk. The weighted country-risk for a raw material or an intermediary product is derived from the individual risk estimates for each of the source countries, which is based on the World Governance Index. The risk value is then determined according to the respective proportion of the global production. The risk includes, among other items, an evaluation of the political stability in the source country and the level of corruption in the country. The higher the HHI value and/or the country-risk value the higher the risk for a reliable supply of the raw material.

China is an interesting example. On the one hand the demand in China for raw materials has increased enormously since the beginning of the millennium. On the other hand, China is also a dominant supplier of raw materials and in many cases the country is the world's biggest producer, for example lead, zinc and tin of the base metals, manganese and molybdenum of the steel alloy metals, or phosphate as a fertilizer. The production of nine of the raw materials that the EU has determined to be critical, the EU-14 or the EU-20, is highly concentrated in China—these are antimony, fluorite, gallium, graphite, germanium, indium, magnesium and the rare-earth elements.

Supply of Raw Materials from Primary and Secondary Sources

Raw Materials are sourced by the primary mining of mineral resources from the geosphere and by the secondary recovery by recycling from the technosphere. The secondary recovery offers significant potential, not yet fully recognized, for raw materials, but it cannot meet the full demand on its own because the requirements are always increasing. Therefore, most of our demand for raw

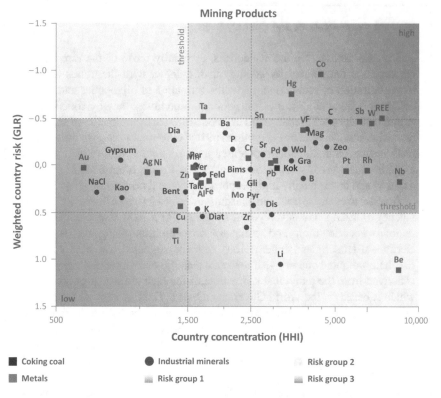

Fig. 3.15 Supply reliability for selected raw materials [58, p. 20]. The selected raw materials can be subdivided into three classes of risk: green represents a low, yellow a moderate, and red a high level of supply risk

materials must today be covered by mining, and this situation will most probably continue in the future.

The location of primary mineral deposits depends primarily on the geological circumstances, and therefore they tend to concentrate into specific regions. Secondary deposits, in contrast, are found most in the urban areas, where most of the waste material and scrap is collected.

Mining companies tend to increase their net worth by take-overs, as is the general case in the business world, because this results in a larger business and cost synergies related to achieving positive economies of scale. Junior exploration companies often undertake the necessary preliminary work, but the construction of a new mine is very cost intensive. The global production therefore may be concentrated into just a few companies. The recycling business, on the other hand, has a pyramidal structure. Several thousand companies and individuals may be involved in collecting the waste materials and scrap, but as the level of processing increases the number involved decreases. In the final stage of smelting (metallurgical extraction and purification of the metals) there are only a few businesses with the necessary metallurgical capacity and understanding to treat complex materials such as electronic parts, batteries and catalysts. In the final stages of processing recycled material, the recycling industry demonstrates similar trends of concentration as in the mining industry.

Natural resources have a geostrategic relevance, and an excessive concentration can endanger the global flow of raw materials. For example, producer countries can cause a supply shortage by imposing export taxes, and this results in premium prices. Similarly, low wages or insufficient social and environmental standards in certain countries can also distort the market competition.

With respect to ensuring a reliable supply of raw materials, it is important to monitor any trends towards a concentration effect that could endanger the free movement of raw materials in the world. Because of this situation, the federal government has commissioned the German Mineral Resources Agency (DERA) to monitor critical raw materials and regularly report on the availability of natural resources that are critical to the German industry.

The industry is also encouraged to secure its supply of raw materials by reacting to warning signals within the necessary time-scale with alternative and flexible strategies and protection concepts. Alternative strategies include, for example, using other raw materials (direct substitution) or technologies (technological substitution), increasing the material efficiency and recycling, diversification of the sources for raw materials, or sufficiently large stocks. These measures can be supported through government by concluding trade agreements, subsidizing research, or financial guarantees. An active role for the state is one option, such as is perceived for the national oil reserves in Germany and other OECD countries.

3.4.2 Political and Social Impacts on the Supply of Primary Raw Materials

With respect to the supply of raw materials, in the ideal natural resource world[12] mineral raw materials would be produced according to the same environmental and social standards, and would be shipped, in any refined form, unimpeded from the

---

[12] An "ideal natural resource world" can be viewed in a broad sense, for example according to the demands of the Federal Parliament Inquiry Commission "Protection of Mankind and the Environment" of 1998 [59]: "non-renewable resources should only be used to the extent that a physical and functional equivalent substitute is created in the form of renewable resources or by increased productivity of the renewable or non-renewable resources". The way in which these demands can be implemented in the market economy is discussed by, for example, Wagner/Wellmer [60].

producer to the markets. Despite all the political efforts to harmonize standards globally and remove trade barriers, especially by the World Trade Organization, we are still a long way from this ideal natural resource world and are likely to remain so in the foreseeable future.

Political Impacts

Political interference is essentially caused by the aspirations of the emerging and developing countries to generate increased wealth in their countries, and they use export quotas and export taxes on unprocessed resources as a lever to achieve this [55, p. 21f]. The example of China and the rare-earth elements illustrates this type of distortion of market competition (Fig. 3.16): since 1993 the established competitors have been excluded from the market by low prices for the rare-earth elements. Once China had established itself as a virtual monopoly, in 2006 China imposed export quotas that resulted in the prices on the world market soaring for selected elements by factors of 100 in 2011. Approximately 75% of the market for finished products containing rare-earth elements, such as permanent magnets, is today controlled by China [62]. This marginalization of primary or secondary production may also result from different levels of the environmental and social standards that are applied to exploitation of these raw materials. Producers must vigilantly make every effort during the exploitation and processing of primary and secondary raw materials to minimize harmful emissions, such as sulfur dioxide, heavy metals or radioactive elements, which occur with the rare-earth elements. This is especially the case if these raw materials are used for the production process for "green energy", because environmental hazards and impacts at the beginning of the production chain would question the credibility of the "Green Economy". However, if higher levels of hazardous emissions and lower social standards are accepted, then the production costs of these raw materials and their intermediary products can be significantly lower. For example, the closure of the rare-earth element mine at Mountain Pass, California, in 2002 was partly due to the strengthening of the environmental conditions for the operation, and the rare-earth element production in California was therefore no longer economic because of the Chinese price-dumping activities. The current high prices for rare-earth elements are therefore partly due to the differences in the environmental standards that are applied in different countries. The market for secondary recovery of raw materials is also distorted in a similar way because many recycling businesses in the developing and emerging countries do not have to observe the environmental and social standards that apply to European businesses. However, these standards are important and the removal of trade restrictions should not be detrimental to the environmentally and socially acceptable production of raw materials. A truly sustainable energy transition can only be implemented if the customers of these raw materials, throughout their value-added chain up to the end-user, insist that these standards are maintained worldwide.

Price Analysis of Intermediary Products

The raw materials themselves should not only be taken into consideration for a risk analysis related to the reliability of supply at acceptable price levels, it is just as

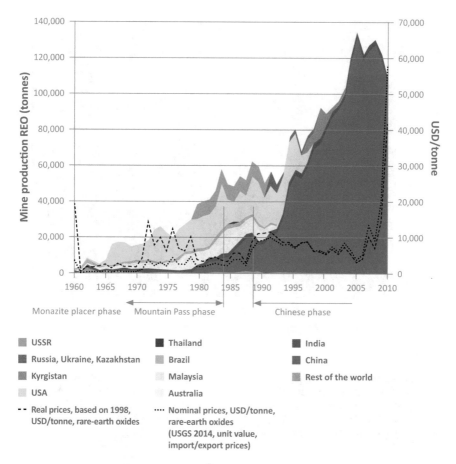

Fig. 3.16 Mine Production (left scale) and Price Movement (right scale) of the rare-earth oxides (REO) in selected Countries (Liedtke/Elsner [61]; updated from BGR [6]). The dashed line shows the real price based on 1998, and the dotted line shows the nominal price in US dollar per tonne (right scale). The major producing countries of rare-earth elements have changed during this time. China began to become a significant producer from the mid-1980s. Initially the global production was mainly based on deposits of monazite (mineral containing rare-earth elements), and later the USA became the principal producer with production from the Mountain Pass mine, California, that is based on the bastnaesite mineral. Since the mid-1990s China has established a virtual monopoly in the rare-earth element market (since the different production phases often overlap each other, the durations of the peak production for each phase are shown beneath the graph)

important to evaluate the prices and supply risks of the intermediary products. This is illustrated by a risk analysis for tungsten that was prepared by the German Mineral Resources Agency (DERA) (Fig. 3.17). There are no longer any export-quotas for the export of unrefined tungsten ores and tungsten concentrates from China. However, the concentration of the production of the highest levels of processed material (ferro-tungsten, for example) and intermediary products with higher degrees of refining

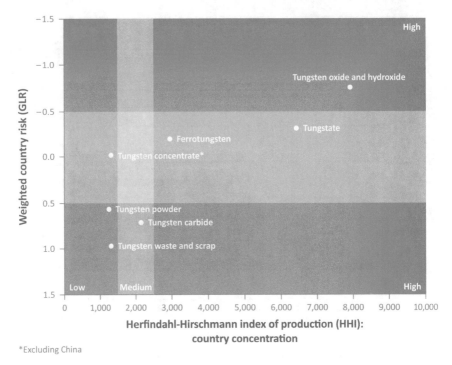

*Excluding China

Fig. 3.17 **Variation in the supply risk along the value-added chain exemplified by tungsten**
[58]. The different levels of processing and refining in the production of tungsten demonstrate,
for the global exports of tungsten, different values for the country concentration and the weighted
country-risk. With respect to the supply of the raw material, the higher levels of processed and
refined product tend to be associated with the higher risks

(tungsten oxide, tungsten hydroxide) is increasing, and they have high HHI values.
These products are mostly derived from countries with a higher risk level, and thus
with lower or even negative weighted country-risk values. As a result, there are
sometimes high price and supply risks: for example, 91% of tungstate is produced
in China, whereas 52% of tungsten oxide is produced in Taiwan and 47% in China
(modified and updated from Liedtke/Schmidt [63, p. 59]).

It should be emphasized that the supply risks for some of the metallic natural
resources are often significantly greater than for the fossil energy sources such as
crude oil and natural gas: the Herfindahl-Hirschman-Index for oil is about 2000
(medial concentration of supply[13]), and for many of the critical mineral natural
resource that are required for the energy transition this value is significantly greater
than 2500 (from this level the concentration of supply is classified as high; see
Fig. 3.17). This means that their production has a much higher concentration of
supply as compared to that of crude oil.

---

[13]OPEC is treated as one country for this calculation.

An industry that is dependent on these raw materials can only protect itself from market distortions of the sort described above by developing suitable precautionary concepts and alternative strategies. These can include, for example, adequate stock-piling, diversification of the sources of supply and strategies to improve flexibility, such as the development of possible substitute materials, so that the industry can continue without, or with reduced amounts of, the specific raw material commodity. In Germany and other OECD countries, for example, the provision of sufficient crude oil (but not natural gas) is regarded as a state responsibility (state crude oil reserves—not only crude oil, but including oil products).

Although the national governmental and international (EU; WTO) counter-measures to remove these distortions to the competition in the markets may perhaps be effective, they are generally neither urgent nor quick.[14] An analysis for the US American energy industry that investigated alternative strategies for critical natural resources described exemplary solutions for the use of rare-earth elements in catalysts for oil processing [66].

BOX V: Borate—Critical Raw Material of the Future?

Borate provides an indication of how a material could develop into a critical natural resource, although borate is currently a raw material without any political significance. Nonetheless, it is interesting because borate could replace antimony as a flame retardant. Antimony, due of its high concentration in China, belongs to the EU-14/EU-20 critical raw materials and the industry has been looking for possible substitutes. Borate could be included as a candidate substitute [67]. Two countries are by far the largest producers of borate—Turkey (more than fifty percent) and the USA. The Turkish government considers that borate is a strategic resource and there are appropriate controls in place. In the current scenario, it is presumed that new energy technologies, for which borate is critically important, will be developed by 2050 and therefore, in contrast to the current situation, borate will become politically relevant. The producer country will then have significantly more powerful economic and political opportunities as compared to the current situation. This exemplary scenario is by no means unrealistic. The use of permanent magnets (iron-boron-neodymium magnets) is currently increasing in the renewable energy technologies [68], and the related demand for borates

---

[14]The complaints about the trade restrictions imposed by China on several mineral commodities that have been recently resolved by the WTO are noted here as an example. On the one hand this was an arbitration procedure pertaining to the rare-earth elements, tungsten and molybdenum that lasted from March 2012 to May 2015 (WTO [64] Dispute Settlements DS431, DS432, DS433). On the other hand, a procedure from June 2009 to December 2012 concerning numerous natural resource commodities (for example, yellow phosphorus, magnesium, manganese and zinc) for which China is the main producing country (WTO [65], Dispute Settlement DS394, DS395, DS398).

is also increasing, so that in future it is possible that borate develops into a
critical raw material.

Social Impacts

There are other uncertainties about the future security of supply of the natural
resources, other than the political issues that result in the tendencies towards concentration that are typical for raw materials. These include above all the social and
ecological aspects.

An essential issue is the fact that the population in Germany, and also in other
countries, has an ever-decreasing understanding for the necessity to mine raw materials. This is primarily because, now that most mineral resources are mined abroad,
the people in these countries have lost their relationship to the exploitation of their
mineral endowment. Furthermore, the applications of the various raw materials are
no longer so well-defined as before. It is no longer so apparent to outsiders why and
where specific natural resources are required or processed. The use of components
in products is today too fragmented and complex. Everyone belonging to a technological society understands industrial products such as televisions, motor vehicles or
smartphones, but only a few know the uses of zinc, tungsten or antimony and how
much of these raw materials, and in what products, they are used. Mineral exploitation
is for many people, particularly those from industrially advanced countries such as
Germany, often an abstract and seemingly normal service industry that is not directly
related to the affluence and technological progress that is desired by the society.

In the resource-rich countries where natural resources are mined, on the other
hand, it is often difficult to communicate the effects of mining. The indigenous population is often affected, although they themselves might never use the raw materials
that are being mined. For example, in Mongolia the veneration of nature is very
important to the culture of the nomadic people. In Australia, the life of the aborigine
people is affected by uranium mine—and that even though the country possesses no
atomic energy plants to use the uranium, which is only mined for export.

The acceptance, or rejection, by the population of mining in their own country
depends on many factors, including the economic development stage, economic
dependence on mining, foreign exchange and tax income, jobs and infrastructure
development that is generated by the mining. The exploitation of raw materials can
have both advantages and disadvantages for the people of a resource-rich country.
The conflicts of interest that are often associated with mining vary from country to
country, and are mostly very complex. A socially and environmentally acceptable
sustainable exploitation of natural resources can only be undertaken if all the various
interests are taken into consideration. In the search for generally acceptable solutions
the various issues should be discussed, and as necessary evaluated, in an open and
objective decision-making process.

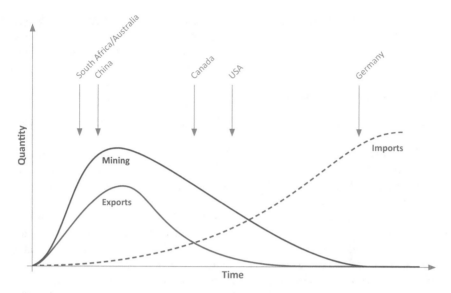

Fig. 3.18 Development from a natural resource producer to an import-dependent industrial country (developed after a concept of Skinner [20]). The various phases of the different countries are shown exemplarily

From Resource Country to Service Economy

All of today's industrial countries were originally also mining countries producing their own raw materials (Fig. 3.18). Important industrial countries such as Germany, France or Great Britain were in the past significant mining countries. During the industrialization process the production factors shifted from primary production (mining, agriculture, forestry, fisheries) increasingly towards the secondary industrial sector, and then further into the tertiary service sector. This is also reflected by the material intensity (see Sect. 3.3.1): the proportion that mining contributes to the gross domestic product continuously decreases.

At the same time the mining industry loses importance. It is telling that Germany and other consumer countries no longer have a specific Ministry of Mines, but the responsibility for mineral exploitation belongs to the departments for economics. In contrast, mining countries such as Kazakhstan, Namibia or Bolivia as well as Canada, which is one of the most important resource-rich industrial countries, have their own Ministries for Natural Resources.

The same trend is found in the understanding of mining: the understanding of the importance of natural resources is much more widespread in Australia or Canada than in Europe. The press regularly reports on mining companies and resource problems. The weekly trade journal *Northern Miner* in Canada reports about mining and exploration companies, and is widely popular especially in northern Canada. Speculation on the so-called penny stocks of the small exploration companies is very widespread (national sport) in Canada, where the Toronto Stock Exchange is the biggest in the world for mining and exploration companies, and Australia, and can be compared the national lottery in Germany. The people are hoping for a major discovery and the associated extreme rises of the share prices, and therefore they carefully follow the daily press reports of the activities of these companies.

Natural Resources—Taken for Granted?

There is generally little interest in Germany for natural resource issues, with the possible exception of oil. On the other hand, books and newspaper articles on the scarcity of natural resources repeatedly arouse significant interest. After China imposed export restrictions on rare-earth elements with the result that the price soared upwards in 2011, even non-experts in Germany became familiar with the names of the rare-earth elements neodymium and dysprosium that are used in the manufacture of permanent magnets in wind energy facilities. Changes in the global supply of oil because of import embargos or military conflicts can quickly result in rises in the gasoline prices, and this also leads to brief periods of public discussion. The anxiety that is generated about the coming scarcity of natural resources is actually an indication of how far-removed the people in Germany, and other countries, are from the real problems pertaining to the mining and supply of these natural resources. Only these risks and environmental impacts are usually the issues at the top of the list of grievances. Consequently, it is rarely understood that it is necessary and rational, even subject to the strictest environmental standards, to exploit primary natural resources in one's own country, which also results in an important diversification of the supply of raw materials. This also has an important knock-on effect in that Germany would set a good example for the future and apply strict environmental standards to foreign companies.

Many people associate the three Ds with mining—"dark, dirty and dangerous" [69, p. 109]. It can be said that mining is associated with a generally negative image.[15] Many people associate mining with not only environmental destruction, but also the danger for the miners. This image is generated throughout the population by spectacular mining accidents. For example, the Lengede accident in Germany in October 1963, which was caused by the inflow from a tailings dam into the iron-ore mine workings, is remembered by many people even today. A few years ago, the subsequent rescue of eleven miners, after 14 days underground, through a wide-diameter bore hole, was filmed as the "Wonder of Lengede". One of the best known recent mining accidents occurred in August 2010 because of a collapse in the St Jose mine in northern Chile. Thirty-three miners were trapped underground, and were first rescued after 69 days also by means of a wide-diameter bore hole. Nonetheless, these events should not conceal the fact that mining activities in the western industrial nations is relatively safe: the accident-rate in the German mining industry is lower than the average in the industrial economy [71, 72]. The high priority given to accident prevention and safety is definitely not the case everywhere in the world,

---

[15]For example, this is the conclusion of Moody [70, p. 11]: "mining always has negative consequences for the local communities, their livelihood and their biosphere".

especially not in the mid-size, small or particularly the artisanal[16] mining operations in the developing countries. In addition to the stereotype image about the dangers of mining, the extensive reporting in the media about environmental impacts also contribute to the negative image of mining—particularly in Germany. Reports give broad coverage to environmental catastrophes, such as the breaching of the tailings dam at the Los Frailes lead-zinc mine (Aznacollar, Spain) in April 1998. The cyanide contamination from the Baia Mare gold mining project, Romania, in January 2000 was also given extensive coverage. The current debate about fracking demonstrates how latent anxieties can be aroused by, for example, oil and gas drilling, because most people have a poor understanding of exactly what happens in the subsurface. Despite what is suggested by the media reporting, only a very small faction of mining projects lead to significant environmental damages.

Impact on the Landscape

Germany is the largest producer of lignite in the world, and is also virtually self-sufficient in the production of construction materials, and even so the impact of this mining and quarrying on the landscape is minor. The total area required in the medium- to long-term to protect Germany's requirements for natural resources is estimated to be less than one percent of the total land area. The surface area equivalent required to produce the natural resources used in 2013 is about 25.2 m^2, which means that about 0.007% of the total area is used for exploitation of raw materials. In contrast to areas used for urbanization and traffic routes, the areas used for production of raw materials are not required permanently and, after a few decades, they can be returned to the community to use as they wish [2].

Obviously, every mining project results in an impact on the landscape, even for those raw materials such as oil and natural gas that require very limited areas of land for production from boreholes. However, major environmental damages derived from the exploitation of natural resources remain an exception. It is important to differentiate between underground mining, by which the natural resources are exploited underground, and surface open pit mining or quarrying, by which the natural resources are exploited directly from the surface.

Underground mining requires relatively small areas on the surface, generally comprising the areas for the technical facilities, such as shafts and processing plant, and the ponds for tailings from the processing plant. These plant facilities look similar to other industrial facilities and, furthermore, no additional changes to the surface facilities, visible to the non-expert, would be expected for the duration of the underground mining. There are outstanding examples for the sensitive adaptation of mine facilities with the local landscape that are fully accepted by the local community [73–75]. Examples include the tungsten mining in the High Tauern region of Austria, or the fluorite-barite Clara mine at near Wolfach in the Black Forest area. Both examples demonstrate how mining can operate with a minimal impact on the local environment in sensitive scenic landscapes. The protection of the landscape is also

---

[16]Artisanal mining is small-scale mining without industrial technology. The mining and processing is manual usually only with the simplest mechanical technology, and rarely with machinery.

observed for the processing of natural resources, for example by processing the raw material in an established industrial zone rather than directly at the mine site.

The situation for open pit mining and quarrying is quite different. The changes to the landscape caused by the exploitation of lignite are clearly visible to people on a daily basis. Villages must be relocated, and the required surface areas may not be available for another use for decades. The long-term effects of open pit mining, such as the effect on the ground water structure, are often difficult to evaluate. These impacts are often not tolerated by the local people, as is the loss of their "homeland" that is emotionally very sensitive to most people. The lignite mining is also rejected by many people because the energy produced is not compatible with climate protection issues. Although in their decision on open pit mining of lignite from December 2013 [76], the Federal Constitutional Court emphasized its importance for the reliability of supply, and consequently the importance of the domestic mining in Germany, but the overall global trend to reject mining of natural resources cannot be denied. The lignite mining areas must be restored, and this can result in attractive landscapes with new lakes. However, the impacts on the environment and local community structures, caused by the temporary exploitation from the surface, are together very substantial.

This negative attitude is not only present in densely populated countries such as Germany and other Central and West European countries, but even in mining countries such as Canada and Australia there is an increasing resistance to mining developments. In some cases this can be identified during the exploration for natural resources, even in the classical mining countries. For example, in the 1970s about one percent of the total area of Quebec Province, where most of the population lives in the south close to the St Lawrence River and the north is virtually unpopulated, was excluded from exploration. In 2011, for various reasons, this figure had risen to nearly 18% [77].

The resistance to mining is most commonly related to open-pit mining projects—particularly in areas with an indigenous community. This raises the question if the current trend to convert underground mines into increasingly large open pits, which mine planners use to take advantage of the so-called economies of scale,[17] is actually the wrong direction. As already described, the acceptance for mining projects depends on several different factors and varies from place to place, and this is demonstrated by the examples of the "underground mine to open pit" trend. Several underground mines were consolidated into one open-pit, the so-called Super Pit goldmine, in the Kalgoorlie gold mining district, West Australia. The local community is long-established but has always lived from mining, and therefore mining is fully accepted. A conversion of the Olympic Dam copper-gold-uranium underground mine in South Australia into a giant open pit has been postponed, despite the area being virtually uninhabited prior to the mining. This demonstrates that such a change-of-opinion is possible even in major mining countries [78]. The protests

---

[17]The "economies of scale" reflects the profit-focused industrial development. By working on increasing large units the fixed costs and therefore the total operating costs can be reduced. The equipment used in open pit mining is becoming increasingly powerful and more efficient (the trucks used today have a 400 tonne capacity, whereas thirty years ago the largest trucks had a capacity of 150 tonnes). As a result, open pit mines can be rationalized to the larger equipment.

against the coal opencast mines in Queensland, Australia, and the associated harbor developments required for coal export, indicate that new mining projects are not always welcome even in classical mining countries such as Australia. In this respect, an interesting alternative is provided by the developments in Austria where open pits are being converted into underground mines because of environmental considerations [79]. As a result, the natural resources can be more selectively mined without impacting the landscape. This can counteract the negative environmental image and maintain, and even increase, the acceptance for mining in the community. The transition to underground mining from open-pit surface operations could be supported by the development of more powerful and efficient underground equipment.

Fundamentally, the acceptance of mining can be increased by awareness-raising measures, inclusion of the community, and innovations. Those who can credibly communicate to the community again that natural resources are a basic requirement for every society will be able to break the negative attitudes.

Opposition in the Developing and Emerging Countries

The acceptance issue has not been given sufficient attention in the past and there are numerous protests in the resource-rich emerging and developing countries, which are often very dependent on the exploitation of their natural resources. The errors of the previous mining are all too obvious. Mining has resulted in broken social structures and contaminated sites, as can be seen in many places in the classical mining countries in South America (Bolivia, Chile and Peru) [80]. The Catholic aid organization, Misereor, is already signaling the foreseeable problems of the planned Tampakan copper-gold mine in the Philippines [81], where about 5000 people must be relocated, most of whom are members of the indigenous population. This demonstrates that social conflict can occur even during the exploration phase. Human rights violations are possible during all phases of the mining process—from exploration to recultivation after the mine has been decommissioned.[18]

Many cases of mining in developing and emerging countries result in the creation of small islands of economic activity, which cement or even enlarge the social differences in the overall population, without positively influencing the total development of a region or a country [80]. As a result of these grievances, the major international mining companies have obligated themselves to Corporate Social Responsibility (CSR—today better known as Social Investment SI) and to environmental standards.[19] They work with recognized international standards, such as those of the International Council of Mining & Metals (ICMM) [83], which is a major international initiative supported by mining companies from several countries. According to the ICMM, 30–40% of the global production is produced by their member companies.[20] The reporting requirements of mining companies are also subject to an

---

[18]BGR (2016-1).

[19]Commdev.org [82]: An overview of the relevant social standards in mining that are internationally recognized, such as for example the International Finance Corporation (IFC), International Labor Organization (ILO).

[20]This proportion should be treated with caution, since China is the biggest mining producer in the world, and no Chinese companies are members of ICMM.

international standard—the Global Reporting Initiative (GRI) [84]. The GRI is an independent, international organization that was formed in 1997 with the participation of the environmental program of the United Nations (UNEP). It aims to promote an increase in the sustainability of economic processes, and undertakes educational work during its evaluation campaigns on economic processes. The GRI has a separate reporting sector for mining—the Mining and Metals Sector Supplement. Despite these efforts, there are still problems. Disproportionately more of the environmental damages are caused by mid-sized or small mining companies that only contribute a small proportion to the global production. These companies also often do not abide by the CSR or ICMM regulations.

High Standards for Mining in the Future

The ICMM was founded in 2001 to further increase the environmental and social standards in the mining sector. The Intergovernmental Forum on Mining, Minerals, Metals and Sustainable Development (IGF) [85], which is co-financed by South Africa and Canada, must also be mentioned in this context. Its objective is to promote a "global dialogue" by which the resolutions relevant to the mining industry of the World Sustainability Conference in Johannesburg should be implemented. The implementation of global social and environmental standards is an additional objective of the IGF. The IGF membership has grown from 25 nations at the time for founding in 2005, to currently 52 nations.

Obtaining acceptance for mining, the so-called "Social License to Operate",[21] is a major challenge to the industry in view of the necessary impacts on the landscape as well as the reservations among the local population against the exploitation of natural resources. However, it is necessary for legitimate primary and secondary raw material exploitation operations. The necessary improvements to the environmental and social standards to be observed by the mining sector is a major, if not the major, task for the mining industry. This is also important with respect to the raw material supply for the new energy systems of the future. Globally Certified Trading Chains (CTC), which are based on transparent, understandable and ethically acceptable production and trading, are one example for the application of such standards in practice, although the smelters have a special role because of their bottleneck position in the chain [88]. The supply chains normally consist of numerous mining operators and numerous customers, but only relatively few smelters (Fig. 3.19) [89]. The introduction and establishment of these certification systems include all the facets of the exploitation of mineral resources and often takes many years. Establishing transparency is a basic requirement to achieve public acceptance.

International banks assume an important function in the implementation of these standards. The banks can insist that these standards are maintained by relevant commitments in the financing agreements. Major mining projects are generally financed

---

[21]Prno [86]: The term "Social License to Operate" was first introduced in the late 1990s by the Canadian Jim Cooney, one of the leading managers in the mining industry.

EY [87]: In their annual risk analysis of the natural resource industry, the consulting company EY (previously Ernst and Young) classified the "Social License to Operate" in the previous two years as one of the five major risks.

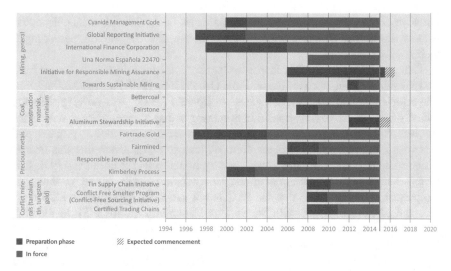

Fig. 3.19 Responsibilities in the Supply Chains [88]. By means of the certifications and commitments in the supply chains, the industry will increasingly ensure a sustainable as well as socially and environmentally acceptable production

by about one-third equity capital and two-thirds loan capital, usually from a consortium of banks. In order to reduce the country-risk of mining projects in developing and emerging countries, an organization such as, for example, the International Finance Corporation (IFC) of the World Bank Group may be included in the banking consortium. Here the Equator Principles [90] must be mentioned as they comprise an international framework for evaluating and minimizing credit risks. The environmental and social standards are conditional for the project financing, and they are based on the environmental, health and safety standards of the World Bank as well as the sustainable operating standards of the IFC.

Despite these efforts, some mining projects still have problems and, in many countries, this is often caused by the high revenues derived from export of natural resources resulting in an increase of the wage and exchange rate levels. This in turn causes new trade barriers in other industrial sectors, and this well-known phenomenon is known as the "Dutch Disease" (BOX VI). There are additional problems in those countries where corruption in the state organizations is rife. As a result, the tax income from the mining sector does not reach the communities, especially in the mining regions. Many resource-rich countries are therefore ranked low down in the Corruption Perception Index (CPI) from Transparency International.[22] In an attempt to address this problem, "Good Governance" [92] sections were initiated by the United Kingdom with the international Extractive Industries Transparency Initiative (EITI) at the World Sustainability Summit in Johannesburg in 2002. The CONNEX

---

[22]Transparency Deutschland [91]: Corruption Perception Index for 2013, ranking table.

initiative of the G7 countries[23] in 2014 should also be mentioned, as its objective is to assist developing countries with respect to the complex negotiations pertaining to their natural resources.[24]

The concept of the "Social License to Operate" has been critically discussed from a social-sciences perspective (for example Owen/Kemp [94]; Parsons/Moffat [95]). The criticism is that the "Social License to Operate (SLO)" was developed by the industry in response to the opposition to mining and as a mechanism to ensure the survival of the mining sector. It is not a real concrete license that is granted, and it is vague and intangible. It is therefore difficult to demonstrate that this license is nonexistent, but at the same time it is easy to claim that it does exist [95]. It is therefore recommended that mining companies should assume at the beginning of the project planning phase that the SLO is not available, and that the SLO must be acquired after dialogue with the various community groups [95]. In the end, the social acceptance of natural resource exploitation can only be realized if the communities are convinced that their values are being respected, the environmental impacts minimized, and the economic advantages are reflected in jobs and better infrastructure for the community.

BOX VI: "Dutch Disease"

The so-called Dutch Disease is a situation for which there is to date no satisfactory solution. It occurs in resource-rich countries, and is especially common in developing countries [96]. The term was coined by the British magazine "The Economist" [97], and describes the effect of a decline of manufacturing industry if an outstanding deposit of natural resources is developed—in this case it was giant Groningen gas field in The Netherlands. Higher wages as compared to other sectors are paid to attract highly qualified workers, and high foreign exchange surpluses were received by the export of the gas. This results in significant fluctuations in the exchange rates because the national currency appreciates in value, and therefore the domestic manufacturing industry is disadvantaged because the price of their products increases for the foreign customers. This effect can be observed today in Australia, which is the world's biggest exporter of iron ore and other raw materials. Automobile production

---

[23]G7 countries are the group of seven countries that together are the seven most important industrial countries on the world. The heads of state and government meet regularly at summits at which numerous current and foreign affairs related issues are discussed. The group includes France, Italy, Japan, the USA, Canada, the United Kingdom and Germany. The group was expanded to include Russia to the G8 in 1998. However, Russia remains excluded from this group since early 2014 because of the Ukraine conflict.

[24]G7 [93]: Point 18 in the communique from the G7 summit of 5 June 2014 includes: "a new initiative of Strengthening Assistance for Complex Contract Negotiations (CONNEX) to provide developing country partners with extended and concrete expertise for negotiating complex commercial contracts, focusing initially on the extractives sector".

is no longer economic, and the plants are in the process of closing. When compared internationally, the wages and production costs are simply too high. This shift in the economic structures is problematic because the commodity prices fluctuate, so that those countries that do not have a significant manufacturing industrial sector are hit hard if a period of lower prices follows on from a commodity boom. This effect can be cushioned in the industrialized countries, as the following examples demonstrate. In the first phase, investment is increased in education and national infrastructure. In the second phase, Sovereign Wealth Funds are established, for example in Norway the income from oil supports the Norwegian Pension Funds, and the Alberta Heritage Savings Trust Fund from hydrocarbon income in Alberta, Canada. Unfortunately, there are only a few convincing examples from emerging and developing countries, such as Botswana and Chile.

Good Governance

Large deposits of mineral natural resources and an appropriately large mining industry do not always bring prosperity and economic growth. On the contrary, a study by the Fraser Institute [98], which is non-profit organization specialized in the economic aspects of mining and is based in Vancouver, Canada, compared the economic growth rates in 77 countries. The average rate of economic growth from 1970 to 2006 was 1.52%. In 26 countries that are dependent to a significant extent on natural resources and mining, the average rate of economic growth was only 1.23%, as compared to 1.66% in 51 countries that were less dependent on natural resources. The 26 countries that are very dependent on natural resources were classified into three subgroups according to their government administration in the Economic Freedom Indicator.[25] The upper third (good government administration) had a disproportionately high rate of economic growth of 2.62%, and the lower third (poor government administration) had a negative growth of −0.38%. Good governance structures must be in place if a country is to benefit positively from an export-based natural resource production. Compliance to appropriate standards in policies pertaining to mining and natural resources could, for example, be prescribed for a country in the financing by international banks. The international banks and private lenders must therefore accept a special responsibility for social and environmental standards.

Political and Social Impacts on the Availability of Primary Natural Resources
Despite all the political efforts to harmonize standards and remove trade restrictions—the WTO is an important player in the latter—a free flow of raw materials is still a long way off. Especially in the industrialized countries such as Germany, where mining only plays a minor role in the national economy, the

---

[25]Gwartney et al. [99], and current data from the Fraser Institute.

public awareness about the necessity for the exploitation of natural mineral resources is declining. For many people mining is associated with the three D's—dark, dirty and dangerous. Because of these attitudes, the mining industry is on its way into political insignificance.

Mining is often associated with serious consequences for the local population, including resettlement and expropriation. The conflicts of interest vary from case to case, and are often very complex. The economic benefits for the affected communities, environmental protection, and the business objectives—a generally acceptable solution must be found to guarantee a sustainable, social and environmentally acceptable exploitation of the natural resources into the future.

In view of the impact on the natural landscape that mining will cause and the doubts of the local population about the exploitation of primary natural resources, obtaining or renewing the acceptance for mining—the Social License to Operate, and therefore a publicly endorsement the exploitation of primary and secondary raw materials—is a significant issue. The improvement of social and environmental standards in the mining sector is a major challenge for the international mining industry, and possibly the biggest challenge for the future.

Good governance structures must be in place in order to ensure that export-oriented production of natural resources is ultimately a positive strategy for a country. The necessary standards for the mining sector and in the resource policies of a country can, for example, be prescribed during project financing by international banking consortia. The international banks and private lenders must therefore accept a special responsibility for social and environmental standards.

The removal of trade restrictions must not be to the detriment of the environmentally and socially acceptable exploitation of natural resources. A truly sustainable energy transition can only be implemented if the customers of these raw materials, along the total value-added chain as far as the end-user, insist that these standards are maintained throughout the world.

3.4.3 Technologies for Exploration and Exploitation of Mineral Resources

Until now it has been possible to find sufficient new reserves to cover the consumption, even with increasing demand. Sufficient supply to meet the increasing demand in the future is an issue that must be addressed. Experts continuously discuss if the efficiency of exploration is decreasing, which is the ratio of exploration expenditures to the number of newly discovered mineral deposits. Some experts Large is the reference [100] suggest that the increase in exploration expenditures since 2003 coupled with a decrease in the number of newly discovered mineral deposits is evidence for this decrease in exploration efficiency (Fig. 3.20a). Others Schodde [102] are not so pessimistic because the tonnage of by-product metal is also an important economic factor. If the by-products are taken into consideration then there is no reduction in the exploration efficiency (Fig. 3.20b). The difficulty of this extrapolation is that discoveries of new deposits are cyclical and occur phase-delayed, which reflects the rise and fall of exploration expenditures. Because of the massive increase in exploration expenditures and the number of new discoveries during the past ten years, many experts predict an increase in planned production capacity of most raw materials during the next five to ten years [103, 104].

With respect to the energy systems of the future, the following aspects of the exploration business must be taken into consideration:

Scientific research on mineral deposits has decreased globally since the 1980s in the previous millennium. Personnel in the institutes for mineral deposit research have been cut back, not only in Germany but also in the classical natural resource countries such as Canada and Australia. This trend was caused mainly by the long-lasting oversupply of natural resources, therefore a typical buyer's market that is now being reversed by the demand from China. Scientific research into exploration technologies is now being revived throughout the world, as is exemplified by the founding of the Helmholtz Institute Freiberg for Resource Technology in Freiberg (HIF). The BMBF supports the r^4 research program on Raw Materials of Strategic Economic Importance for High-Tech Made in Germany [54] that is focused on, among others, the development of methods for mineral resource exploration.

Classically exploration commences with near-surface investigations. Although the principal methods for surface exploration are well known, they are continuously being refined and upgraded. One special innovative advance during the next few years could be related to airborne hyperspectral analysis, which analyzes different frequency bands that are reflected from the surface of the earth. This data can provide information about the mineralogy of the soil, and it can already map hydrous minerals that might be indicative of the presence of mineral deposits or even rare-earth elements directly. The method is particularly effective in arid and semi-arid areas. The sensitivity of the analytical instrumentation is continuously improving so that the prospectivity of large and unexplored areas can be assessed by satellite. As part of the EnMAP (Environmental Mapping and Analysis Program), from 2018 a new satellite will bring these systems that were exclusively developed in Germany into earth orbit. Aerospace issues are managed by the German Aerospace Center (DLR), and the scientific control is with the Helmholtz German Research Center for Geosciences (GFZ), Potsdam. Other scientific institutes participating in the project include the Remote Sensing Technology Institute (IMF), the German Remote Sensing Data Center (DFD) and the German Space Operations Center all of the DLR, OHB System AG, the Kayser-Threde company, and the Federal Institute for Geosciences and Natural Resources (BGR).

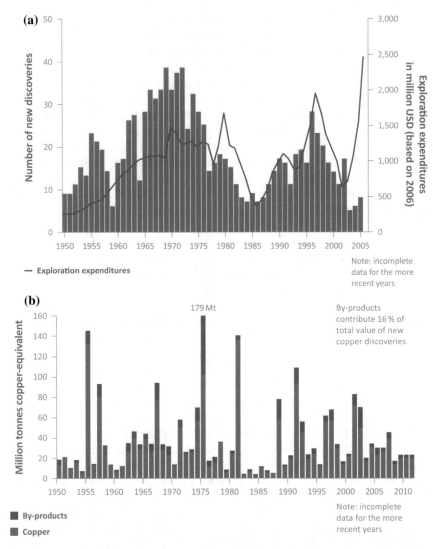

Fig. 3.20   a Exploration expenditures and the number of base metal discoveries (copper, nickel, zinc, lead) in the period 1950–2005 (modified from Large [100]; translated after Bizzi [101]). The exploration expenditures are corrected for inflation. The discoveries include those with more than 0.1 million tonnes of contained metal converted into copper equivalent (see Fig. 3.20b). **b Tonnage of copper and by-products in deposits with greater than 0.1 million tonnes contained copper that were discovered during the period 1950–2011 (presented in million tonnes copper-equivalent, Cu-eq.)** [102]. Estimate includes adjustments for deposits with no discovery year and deposits missing from the database. The proportion of the by-products is calculated on basis of the monetary value of one tonne of copper ore containing 1% copper is equivalent to the value of one tonne ore with 3.26% zinc, 4.76% lead, 0.30% nickel, 0.25% molybdenum, 0.43% cobalt, 0.94 lbs uranium oxide (U_3O_8), 0.44 tonnes magnetite, 3.0 grams per tonne gold, or 156 grams per tonne silver. (*source* Schodde [102], with permission of the author)

The greatest potential for discovery of new mineral deposits is at depth: discoveries until now (Fig. 3.21) have been relatively close to the surface, but since about 1990 new orebodies have been discovered at greater depths below surface. This is in part due to the improvements over the last 30–40 years to electromagnetic exploration technologies. In the 1970s helicopter-borne electromagnetic signals used to penetrate to depths of only 100–200 m into the earth's surface, but today they can penetrate three times that depth. During the recent decades, there have no longer been the radical and successful innovations such as the application of airborne magnetic surveys for mineral exploration, which was a method developed during the second World War for submarine detection, or the early airborne electromagnetic technologies that resulted in a new discovery of base metal mineralization almost every year from 1950 to 1976 [106]. However the SQUID magnetometer (based on superconductors), which was developed by the Leibniz Institute for Photonic Technology, Jena, together with a South African mining company, is an example of one successful technical breakthrough. First results suggest that the SQUID, which is based on superconductor technology, can identify structures at even greater depths [107]. The Federal Institute for Geosciences and Natural Resources in Hannover is also working on mounting the so-called AFMAG system (audio frequency electro-magnetic) into a helicopter-borne bird, and expect that this will penetrate to depths of up to 1000 m. Geochemical techniques are also being advanced to detect mineral deposits from chemical trace elements in the surrounding rocks. This requires very sensitive and accurate analytical techniques so that the commonly very low concentrations of the trace elements can be reliably measured.

The exploration for deeply buried ore deposits is expensive. Smaller exploration companies, which are often more creative and operate more cost-effectively than the bigger companies, usually find it difficult to raise the necessary finance. They will therefore lose their importance.

Exploration is a component in the feedback control cycle of raw material supply. If the rate of new discoveries remains behind that of production for a longer period of time, then the market will regard the impending physical shortages as a risk, and the prices of these natural resources will increase. In turn, this will encourage the exploration industry to focus on these resources and develop new exploration strategies. Uranium exploration in the 1970s presents a good example of this: the exploration costs increased with time from 10 US dollars per pound (USD/lbs) for uranium oxide (U_3O_8) to 40 USD/lbs. New methods were developed for exploration with greater depth penetration, particularly to target the rich uranium deposits in geologically favorable locations in Canada. Numerous high-grade deposits were discovered as a result, and this in turn resulted in a price decrease such that the lower-grade deposits became unprofitable.

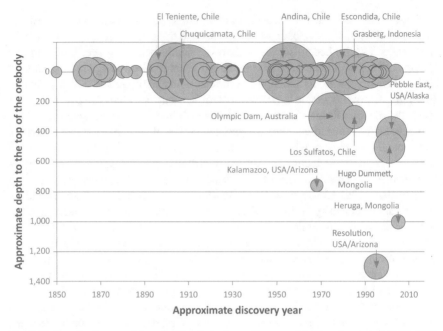

Fig. 3.21 Depth of copper deposits that have been discovered during the period 1850–2010 (modified from Large [100]; translated after Finlayson [105]). Deposits containing more than four million tonnes of copper are included

---

BOX VII Effects of Rationalization—The Example of Copper

The First World War and its consequences resulted in a reorganization of global mining production. Technical advances and resultant rationalization such as the introduction of large-scale open-pit mines (economies of scale) reduced the costs of exploiting natural resources, and this is the reason for the collapse in the price trend after the end of the First World War. The price of copper, for example, has not increased in real terms during the past hundred years, although the copper grades of the deposits that are being exploited have in part decreased substantially (Fig. 3.22) [14]. The increase in size of the equipment used in open pit mines and the improvement in the metallurgical processing of copper ores both contribute to greater efficiency so that deposits with a copper content of only 0.27%, such as the Aitik Mine, Sweden, can be successfully mined [111].

The average number of newly discovered deposits per year has remained approximately the same since 1950 [102]. However, since not all the new deposits are, or were, mineable at the time of their discovery, many of these discoveries remain classified as resources. Technological improvements are the principal trigger for their transition into reserves. For example, about twenty percent of the global production of primary copper is derived from the SX/EW (Solvent Extraction/Electrowinning)[26] process that is nowadays applied to low grade copper ores. This technology has also been successfully applied to the production of cobalt, nickel, zinc and uranium. By-product metals in copper deposits, such as the precious metals or tellurium (a technology raw material), are not recovered by these leaching processes, and this is a distinct disadvantage of the methodology.

The Improved Availability of Natural Resources resulting from Technical Innovation.

The availability of natural resources is also enhanced by technical improvements in the exploitation and processing. Thanks to new technologies, resources, or known deposits that were never economic to mine, can evolve into economically mineable reserves. Furthermore, new low-grade ore deposits that have not yet been discovered can become a focus for exploration activities. The new technologies make it possible to decrease the threshold, or cut-off, for the mineability, and the tonnage of economically mineable reserves increases disproportionately. Price incentives that reflect the feedback control cycle of raw material supply are also important.

Economic rationalization in the mining industry results from increasing the unit size (economies of scale) and from technical advances (BOX VII and BOX VIII). The development of remotely controlled mining equipment is very important. The processing methods are improved by using energy-saving crushing and milling technologies as well as from leaching methods, including bioleaching with bacteria (BOX VII). These developments can also act positively on the economics for the reprocessing of waste materials, such as the tailings from processing, slag heaps and ash from power plants, and thus at least temporarily improve the reserve status of the electronic metals or the rare-earth elements.

BOX VIII: Technological Innovations—The Example of Unconventional Hydrocarbons

The development in the USA of shale gas and shale oil, or the unconventional oil and natural gas resources, clearly illustrates just how radical the consequences of new innovations such as the fracking technologies, which allow previously untouched primary hydrocarbon deposits to be exploited, can be for the supply of natural resources. Fracking has become the standard method

---

[26]The process is a two-stage hydrometallurgical method for extracting the principal commodity (in this case copper) from the ore by means of leaching (solvent extraction) and subsequent electrolysis (electrowinning).

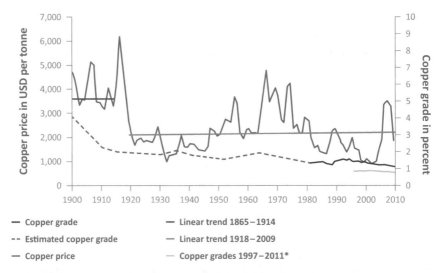

Fig. 3.22 Development of the average global copper grades (Schodde [108]: based on the primary production of copper from milling or leaching processes) and the real price for copper from 1900 to 2010 [27, 28, 109]. The global consequences of the First World War and rationalization resulted in a reduction in the mining costs that is also reflected in the collapse of the price after the end of the war. The two horizontal linear trends correspond to the respective levels of the average prices (sources Schodde [108], Stürmer [27, 28], Scholz and Wellmer [109], with permission of the authors). (*Additional data from 1997 to 2011 that also includes the leaching of very low grade waste dumps[110] includes copper from leaching of waste dumps that typically have a grade of only 0.1–0.3% copper, and therefore this distorts the average grade downwards because this material is not taken into consideration in the development of new mines. This exploitation of the mine waste is not economic in relation to the overall costs of a mining project, but provides the possibility for additional income since the process is economic for such low grades when treated as an "add-on project". This implies that there is currently no necessity to exploit such low grades, not because of the lack of higher grade deposits, but because it is technically and economically possible, and therefore this activity reflects the technical advances in the metallurgical processing (personal communication Schodde 2015 [112])

of hydraulic fracturing, which ruptures underground the less permeable rocks by high fluid pressures, so that the hydrocarbons can be extracted from the sealed pore spaces. In contrast to the conventional deposits, from which the hydrocarbons can be extracted with relatively little technical effort, fracking technologies are required to extract the hydrocarbons from these rocks. Even if some predictions are clearly too optimistic, it is possible that the USA will soon be self-sufficient in natural gas. The gas prices in the USA are already only about thirty percent of those on Europe.

The boom in American shale-gas is the result of significant technological advances that make it possible today to economically extract these quantities of shale-gas. The development time of nearly twenty years for hydraulic rock fracturing (fracking) is shown in Fig. 3.23 as an example of the typically

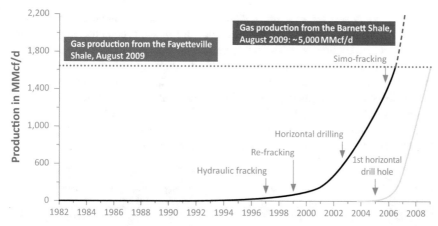

Fig. 3.23 Learning Curve for the Exploration and Development of Technologies, as exemplified by the Production of Shale Gas in the USA [113]. MMcf/d is the unit, Million Cubic Feet per Day, typically used for natural gas production (one cubic foot is approximately 0.02832 m^3)

long learning-curve for technical breakthroughs in the natural resource sector: after the initial development of the fracking technology, the methodology is refined (re-fracking is renewed fracking in an already existing borehole) and with other technical advances (horizontal drilling and simultaneous fracking in several parallel boreholes—simo-fracking). The production from the Fayetteville Shale, which is a geological formation in Arkansas, USA, has benefited from the learning experiences in the Barnett Shale, Texas, where the fracking technologies was tested much earlier.

The energy requirements for the exploitation and processing of mineral natural resources is also relevant from the environmental perspective: mining at greater depths as well as less rich and more complex ores generally requires an increasing energy input (per tonne of metal recovered). Already today about eight percent of the global energy requirement and, respectively, the CO_2 emissions, are directly related to mining [114]. Although these additional costs might be covered by rising commodity prices, but in the worst case the energy consumption for the generation of renewal energy increases significantly because of the disproportionate work input at the beginning of the production chain. At least a proportion of the advantage is therefore used up if it does not prove possible to apply renewable energy in global mining activities.

Marine natural resources are also relevant to the future availability of natural resources for the new energy systems. The Federal Government, through the BGR, has acquired concessions in the Pacific Ocean for polymetallic nodules (previously known as manganese nodules) that contain copper, nickel and cobalt, as well as a concession in the Indian Ocean for massive sulfide ore deposits, containing base-metals, over extinct submarine hydrothermal springs. The massive sulfides also contain by-products that include several electronic and high-technology raw materials. Although the technologies for the exploration of these deposits are well advanced, those for the exploitation in the deep ocean and the processing are still in the early phases of development. This is particularly the case for the polymetallic manganese nodules. The permitting for exploitation licenses and the development of the necessary technologies will probably take at least another ten years.

> **Development of Technologies for the Exploration and the Exploitation for Primary Natural Resources**
> The expertise in Germany required for the exploration of natural resources has mostly dwindled away because of the reduction of mining activities. This includes the scientific, industrial and financial facets. Increases in the efficiency of exploration, production and processing all lead to an enhanced availability for natural resources. The following challenges relate to research and development: the penetration depths, the precision and the surface coverage of exploration must be improved by technical advances and the optimal combination of methods from the fields of geology, geophysics and geochemistry so that "blind" mineral deposits, with no surface expression, can be discovered at depth and unexplored areas can be systematically explored. The mineability of mineral deposits with lower grades and complex relationships should be improved by technological improvements so that they can be economically exploited in the future. The energy balance of the metallurgical recovery processes should be more sustainable. Marine deposits and dumps of waste from mining and processing will also play a role in the long-term.

3.4.4   The Availability of Secondary Resources

The metallic natural resources that are mined in the resource-rich regions of the world are primarily used for products used in the rich industrial countries. So long as reusable materials are recycled, these products represent an important deposit of secondary natural resources at the end of their life-cycle. Not only automobiles or computers belong to the potential secondary deposits, but also infrastructure such as buildings or electricity grids. These secondary deposits are mainly concentrated in the urban areas. If these secondary deposits are comprehensively exploited, then this significantly increases the natural resource basis of the national economy. Recycling is therefore fundamentally a good opportunity to reduce the dependence on the supply of critical raw materials from primary sources. Technically this is possible, because modern metallurgical processes can extract metals from secondary sources that have the same quality, including purity and physical-chemical specifications, as metals from primary sources.[27]

Technology metals must usually be metallurgically extracted from complex fractions; from ores and secondary materials that often contain a mixture of different metallic materials. The pre-processing is usually undertaken regionally, but the metallurgical processing of complex mixtures of metals can only be undertaken in one of a few large metallurgical facilities in, or even outside, Europe (Fig. 3.24). European metallurgical pants process concentrates, which are produced from both primary as well as secondary sources. The pure metals that are extracted by the metallurgical process are traded on the same commodity markets for the same prices, and are used in the same products. Since the use of secondary sources requires a continuous feed

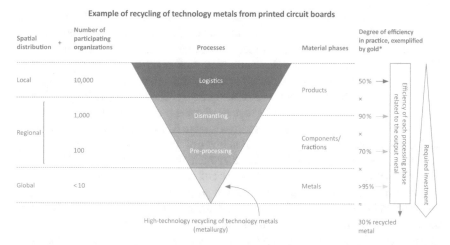

Fig. 3.24 Recycling process chain for consumer goods—the example of recycling of technology metals from circuit boards [115]. The overview demonstrates the complexity of the process chain and the economic challenges associated with this: the logistics level, demonstrating where the secondary raw materials are accumulated and processed, and how many organizations are involved, is compared with the technical levels in the recovery steps (*the effective recycling rate (recycling efficiency) for a metal is derived from the product of the efficiency of each of the steps (from Hagelüken [116, p. 165]). In this example 50% of the relevant waste appliances are collected (collection efficiency), during the processing 30% of the gold is lost to the "false" fractions (steel, plastic, dust, etc.). The numbers used are realistic for the recovery of gold from consumer electronic appliances in Germany and Europe.)

---

[27]Only the electrochemical characteristics of chemically ignoble metals, such as aluminum and magnesium, can suffer a certain down-grading effect. In the case of a mixture of different aluminum alloys, for example, then some of the alloy metals can be recovered only with a very significant energy input. This means that after the simultaneous recycling of different aluminum alloys the resultant alloy is usually not suitable to be used in a low-alloy material ("low-alloy" means that a metallic material contains only low quantities of alloys). It is therefore important, by means of scrap processing and sorting, to avoid these mixing before the smelting.

from the active economic cycle and the technosphere, the inventories can be readily estimated. The quantities in the secondary deposits of natural resources are derived from the number of products that have been sold or the constructed infrastructure and their estimated lifetime or, respectively, operating life (see BOX IX). These used products and scrap materials have a completely different composition as compared to the primary geological deposits. Because the raw materials that are used in the production processes are derived from different types of geological mineral deposits, much more complex "scrap deposits" are created in the technosphere. They contain a much greater variety of elements and compounds that are usually not present in nature. A mobile telephone contains, for example, more than forty different metals. Furthermore, these elements are closely interlinked together with inorganic and organic materials in manufactured products, such as metals and plastics in printed circuit boards or electrical cables, or steel alloys with construction materials in reinforced concrete.

In any specific geographic region, and depending on the standard of living, a variety of very different products are accumulated that differ in their size, complexity and distribution—for example ships, trains, automobiles, refrigerators, computers, cell phones or USB sticks. Some of these products, such as ships, represent an important source of metallic raw materials just because of their size. Other products, such as smart phones, are important because of the quantity. The heterogeneity and also the rapid changes caused by the short life cycles of some products mean that the processing of these deposits of secondary resources is related to very different challenges and subject to changes in space and time. Examples of these changes include, as already described, the development of the flat-screen television from the tube television, or the flash memory from hard disk drives. The effective recycling and comprehensive re-use of these secondary materials requires them to be identified (or "explored"), recorded and presorted. It is just as important to develop an appropriate logistical network. Each of the steps in the treatment chain, the logistics, the dismantling and the mechanical preprocessing; and the final metallurgical recycling processes must be carefully coordinated, as ultimately the level of the targeted recycling rate is dependent on this (Fig. 3.24).

Recycling Rates

High levels of recycling of more than 50% have so far only been effectively achieved for the major and precious metals (Fig. 3.25). In addition to the further development of the metallurgical processes, incompatible compounds of various elements should where possible be avoided by applying appropriate product designs. The supply of secondary resources and the appropriate quality of scrap (including the content of the secondary resources, their physical composition and heterogeneity) are a challenge to the European smelters for the recovery of these raw materials, and they must be continuously customized for this purpose (Table 3.1). The development of new fragmentation and sorting processes is also important, so that certain components or materials that are incompatible to a specific metallurgical process can be sorted out without losing any of the other valuable materials. Pre-metallurgical sorting, such as for example the separation of magnets from electronic appliances, is a prerequisite

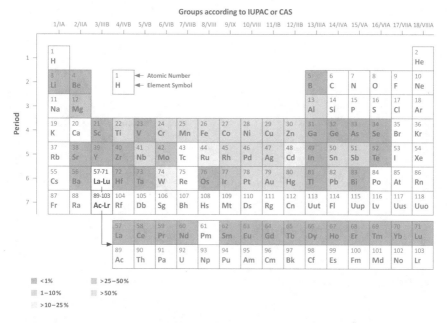

Fig. 3.25 Average global rates for a functional end-of-life recycling of sixty metals (after UNEP [117]). Functional recycling means that the properties of the raw materials—the same as for primary resources—remain for further use in new applications. For the elements that are not color-coded, there is either no data or estimates are not possible, or it was not analyzed in this study. Metal-bearing emissions from coal-fueled power generating stations were not taken into consideration. *Source* UNEP [117], with permission of the United Nations Environment Program UNEP

to enable the recovery of the rare-earth elements. Without this sorting the rare-earth elements would be lost in the copper metallurgical process. On the other hand, for example the cost-effective separation of small tantalum capacitors from circuit boards, is usually not possible without also accepting the partial loss of the palladium and silver that is often contained in the same capacitors and circuit boards.

Because the depositories of natural resources in the human technosphere are continually increasing in size (metals are used and not consumed), the quantity of natural resources that can be recovered from secondary materials will increase with the development of more efficient recycling systems, as is demonstrated by the example of aluminum (Fig. 3.26). According to information from the International Aluminum Institute in Paris, the secondary resources contributed a proportion of 17% to the global aluminum production in 1960, but by 2009 that proportion had risen to 30%, despite a significant increase in the total production, and in 2020 it is predicted that this proportion will rise to 37%. In 2016, 43% of steel production in Germany was derived from secondary sources, and the proportion for aluminum was as much as 57%, 41% for copper [120]. The recycling rates for lead are estimated by experts to be around 60% and for zinc around 20%. The recycling rate is different

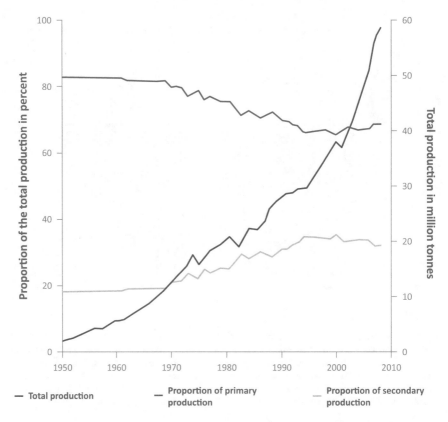

Fig. 3.26 Development of the global production of aluminum from primary and secondary sources (IAI, http://www.world-aluminium.org). (*Source* International Aluminium Institute (IAI) [119], with permission of the IAI)

for each metal, but it is common to all metals that the recycling rate is not sufficient to cover the total demand.

BOX IX: Effect of Residence Times on the Potential for Secondary Resources—the Example of Copper

For the recovery of raw materials, and especially metals, from the secondary deposits in the technosphere, it must be taken into account that the metals all have a residence time in the active economic cycle. The metals are only available for their recovery at the end of the lifetime of the relevant product, which is no longer used in the active cycle. If the demand for raw materials does not stagnate during this residence time but increases, then the quantities available for recycling are always less than those required by the demand.

Table 3.1 Recycled materials used for the recovery of copper in Western Europe, and their copper contents [118]

Formerly: the "classic" recycling materials	
Copper scrap	94–99% copper content
Alloy scrap	50–90% copper content
Residues (slags, dross, dusts)	15–60% copper content
Today: increasing use of the "modern" recycling materials	
Shredder–copper concentrates	25–60% copper content
Computer circuit boards, not mechanically processed	12–16% copper content
WEEE–materials (electronic and electric equipment waste)	4–20% copper content
Industrial catalysts, slimes from industry etc.	1–50% copper content
Processed "landfill-mining" raw materials	>10% copper content

The concept that, during periods of increasing consumption of raw materials, a national economy can live from its secondary raw materials, is clearly not applicable (Fig. 3.27).

This relationship can be clarified with the example of copper. The lifetime of appliances and infrastructure differs significantly between the various sectors—in the electronics sector it is only a few years, but in the construction sector it is 40–50 years. For this conceptual example, an average weighted lifetime of 30 years is presumed for all sectors.

In 1982 the global copper demand was 9.1 million tonnes, and in 2012 it was 20.6 million tonnes. If a global collection efficiency of 90% (effectiveness of the first recycling step) is optimistically assumed, then for a product lifetime of 30 years there would only be 8.2 million tonnes from the copper used in 1982 available for recycling, which is about 40% of the global copper consumption in 2012.

The figures for Germany are rather better because, during the period under consideration, the total demand has not risen so strongly. The consumption in 1982 for the Federal Republic and GDR was 880,000 tonnes and this had risen to 1.1 million tonnes in 2012. Assuming a collection efficiency of 90%, which is probably realistic for Germany, there would theoretically be 790,000 tonnes copper available in 2012 from recycling, which would be 70% of the demand in Germany. In Germany, the actual recycling rate (recycling efficiency) for copper was between 50–60% during the period 2000–2009.

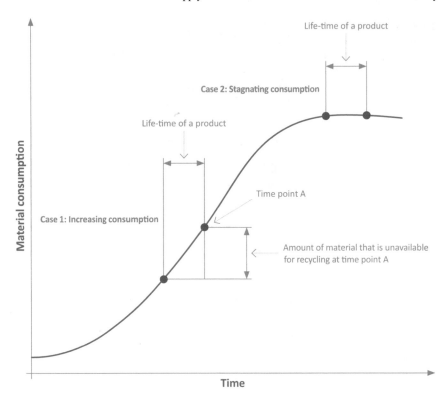

Fig. 3.27 Idealized growth curve for natural resources (modified from Steinbach/Wellmer [121]). The opportunities for recycling are limited. Case 1 shows that during the growth periods always results in a deficit of the proportion of secondary material that is theoretically available later. Only in Case 2, with constant consumption, is the same amount of secondary material available later

Recycling of Complex Mixtures

The ideal case of covering a specific raw material demand to one hundred percent by recycling is not only unattainable because of the product lifetime or, in other words, the residence time, but the complexity of the metal combinations in many products is another limiting factor which constrains that the metal's demand can be met without mining. The "100% recycling" of all elements at a product's end-of-life is not possible for complex combinations of materials. Depending on the selected recycling process, there will always be losses of some elements, as was illustrated by the example of recycling tantalum capacitors on printed circuit boards. Furthermore, in some appliances specific raw materials, often the technology metals, are not present in concentrations greater than in the primary resources, but are only present in very small concentrations. For example, the indium concentration in television glass screens is less than the concentration in sphalerite, which is the most important ore for by-product indium.

In this respect, the energy requirements must also be taken into consideration if the objective of recycling is also to reduce the overall energy and therefore also the environmental impacts. This is a further limiting factor to the utopian ideal to recycle one hundred percent of all natural resources that are in use. Figure 3.28 shows the energy requirement per tonne of metal against the incremental amounts of metal, and these are ranked from left to right according to the energy requirements for the respective recovery. The graph on the left shows separately the energy requirements for primary and secondary resources, and the graph on the right shows the energy required for a mixture of primary and secondary raw materials. In order to meet the demand for a year's consumption of metallic raw materials, different successive sources must be used that have a higher energy requirement, for example a low concentration of the raw material in the resource, or more complex mixtures of materials, and this can be seen from the slope of the relevant curve in both graphs. The increase of energy required for secondary resources is generally more pronounced. Although, as compared to recoveries from primary resources, a higher proportion of material can be recovered from secondary materials with less energy, the required energy input can sometimes be much higher for the worst case such as complex and low grade mixtures of materials. The graph on the right shows that in this example an optimum energy usage is attained with a mixture of about 65% secondary to 35% primary resources. So long as those secondary resources, which can be recovered with more a favorable energy input as compared to that required for the primary resources, are available then the energy input decreases with an increasing proportion of secondary material in the mixture. However, if a very high proportion of secondary resources is required, then it is necessary to input material from which it is very costly to recover the required raw material and, in this case, an increase in the proportion of secondary material will lead to a higher energy requirement.

The secondary recoveries from relatively pure metal are always more energy efficient [121]. The energy saving by using secondary aluminum is about 95%, about 85% for copper, and about 65% for lead. Copper is also an ideal "collector"[28] for precious and some associated metals, such as selenium, tellurium or nickel. However, as the secondary raw materials become more complex, then the energy input rises for the recycling, unless it is technically possible to recover several metals simultaneously from the same complex raw materials. For example, if precious metals are contained in circuit boards, the recovery is generally possible with a positive energy balance, even for low concentrations and complex mixtures, because the concentrations in the primary mineral deposits are often even much lower and because even traces of precious metals "automatically" follow the copper through the smelting process. As described above for the example of indium in television screens, the concentrations in the secondary sources can be even lower than those in the primary deposits. The energy balance is therefore not necessarily better for the secondary resources. The energy requirements for the recovery of raw materials can be minimized by using an optimal mixture of secondary and primary materials (Fig. 3.28, right).

---

[28]The term "collector" means that other metals are collected in the melt, and they must then be separated by other processes.

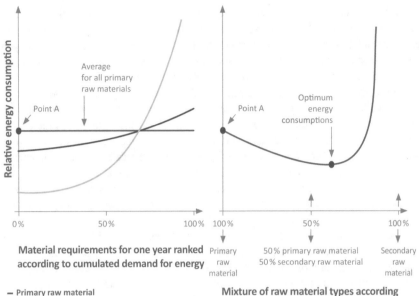

Fig. 3.28   Energy Input for Metal Recovery from Primary and Secondary Resources (modified from Steinbach/Wellmer [121]). Point A is the base-point in both graphs, and shows the average energy input required for the recoveries from primary sources

In the future it can be expected that the energy input required for recovering raw materials from secondary materials will not change very much, but that required for the primary resources will increase because it is to be expected that the mineral deposit characteristics will become more complex. The reasons for this could include the grade of the relevant raw material might be lower in new primary mineral deposits, the ore deposits are located at greater depths, and the ores will become more complex and hence difficult to process. It can therefore be expected that the theoretical optimum for the energy required to recover metallic resources, as shown in Fig. 3.28, will shift towards even higher proportions of secondary materials and less primary material. However, there is still a good possibility for mining to become more energy efficient, as shown by a 2007 study from the US Department of Energy (DOE) [122] on American mineral exploitation (except oil and natural gas), which demonstrated an energy saving potential of 53%.

Optimization of Recycling

In many cases recycling is either non-existent or sub-standard outside Europe, and this leads to an avoidable loss of resources and also impairs the overall energy balance. This situation could be managed by improving the management of the flow of materials as well as the introduction and certification of global recycling standards (BOX X). In the final analysis, the circular economy must be organized in such a way so that once metals have been introduced into the technosphere, then as much as is feasible will be recovered at the end of their lifetime in products and infrastructure. Only so can the maxim "use metals, instead of consuming metals" be implemented in practice by the society.

Metallurgical recovery is the final step in the process chain, and requires complex technology and high investment. Because of this, the metallurgical process in the extraction of both primary and secondary resources is subject to the same economies of scale as in other industrial sectors.

The long-term objective of the resource strategy must be to meet the total resource demand as much as possible by recycling. Once metallic resources have been extracted from the geosphere into technosphere, they should be continuously transferred with minimal losses into new product life cycles (Fig. 3.29, BOX X). This can only be achieved with a step-by-step approach, although in some cases it is already almost the situation today. For example, in Germany the lead required for batteries can be completely covered from secondary sources [121]. Tungsten, which is used for tungsten-carbide tips in metal works, is another example for a life cycle with minimal losses. The tips are re-ground or recycled, and only the ablated tungsten-carbide is replaced.

BOX X: The Problem of Incomplete Material Life Cycles—The Example of Metals

The secondary deposits, except for major infrastructure such as electrical transmission lines, differ from the primary geological deposits in that they are not confined to a fixed location. Typical examples include automobiles, computers, and cell phones. The end of their lifetime does not necessarily mean that these products are available for recycling. For example in Germany and Europe, significant amounts of secondary raw materials are lost as legitimately exported used goods, but are also suspiciously or illegally exported as waste materials. The resources are exported as products, components (for example as printed circuit boards or catalysts) or as processed fractions.

Because in many cases recycling is either non-existent or sub-standard in developing and transition countries outside Europe, this leads not only to an avoidable loss of resources for Germany and Europe, but also to irreversible dissipation from the global circular economy. The size of this loophole between the potential of secondary deposits of resources and the actual secondary production depends mainly on the prevailing technical and social-political conditions. With respect to the latter, the implementation of waste regulations is extremely important. The export of waste appliances results mainly from economic incentives, whereby in many cases the impetus is the externalization of environmental and social costs. The reasons are not better recycling processes in non-European countries.

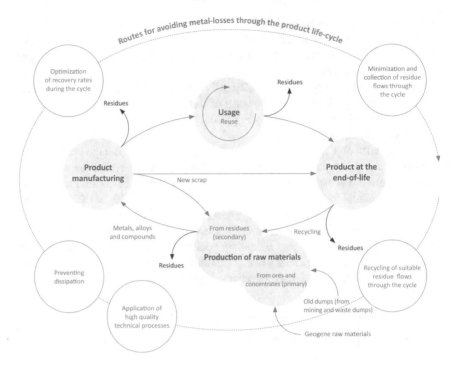

Fig. 3.29 Idealized Circular Economy for Metals (modified after Meskers [122], Hagelüken [116]). The lifetime cycle of metallic resources from the primary mining exploitation to recycling includes losses, short-cuts and interdependencies. Increases in the productivity in the individual stages of the cycle and the avoidance of losses are required to optimize the material cycle and increase the raw material efficiency

Finally, looking forward to 2050, it is predicted that the geo- and technospheres have sufficient potential to meet an increasing demand for resources. This is conditional on a functioning market economy that can react to price incentives (see Sect. 2.4). Additional conditions include an efficient and sustainable infrastructure, both technically and economically, as well as research, whereby the industry—as part of their responsibility for resource supply—will be required to lead projects on resource supply from both primary and secondary sources to full commercialization. The industry can be supported at the outset of this endeavor by, for example, appropriate formulations in the general political conditions or inventive targets. Furthermore, focused support for research from the state would be an additional incentive.

3.4.5 Development of Technologies for the Recycling of Secondary Materials

The "metal wheel" (Fig. 3.30) was developed by metallurgists and provides a good basis for discussing the problems in recycling [125, 124]. It is a presentation that classifies metals according to the methods used for their metallurgical processing. The metallurgical and thermodynamic laws for metals are the same, regardless if they are derived from primary or secondary sources. For example, the oxidation behavior and the metallurgical affinities of metals for each other, or also the melting temperatures, do not change. Metals can therefore be presented on a "metal wheel" regardless of their source, but the association with other metals/elements in ores or in products, for example secondary groups, are important. Depending on the raw material to be recovered and the composition of the source material, there are various metallurgical routes to optimize the recovery process (BOX XI).

The metallurgical recovery of the metals is the final step in the recycling process, so that increases in overall efficiency can only be obtained by an optimization of all the steps in the process chain. There are four significant challenges for the development of measures and technologies for recycling of secondary materials:

1. The total efficiency of the recovery of secondary metals is the product of the recovery rate in each of the individual steps (see the example in Fig. 3.24). The first step is the collection, which is the most important step. Particularly for many of the metals important for the energy systems of the future, the recoveries at this first step are very low. Only a small proportion of consumer electronic appliances, for example, are collected for recycling. The contents of valuable metals in each tonne of electronic scrap are often higher than in one tonne of primary ore[41]: for example, computer printed circuit boards contain 100–200 grams per tonne (g/t) gold and up to 50 g/t palladium, or cell phones contain 300 g/t gold and 40 g/t palladium as compared to 1–10 g/t gold or 1–5 g/t palladium in ores. The specific metal value for cell phones is, however, low (about 1€ per cell phone), so that there is no commercial incentive to finally submit these products for recycling. About twenty percent of global mine production of palladium and cobalt today are used in short-lived consumer appliances such as cell phones, PCs and laptop computers. For both gold and silver this value is four percent. The unused potential for these high-technology and precious metals is therefore estimated to be high, but the utilization of this resource potential depends on overall recycling infrastructure and the motivation of the end-user to submit their appliances for recycling at the end of their useful life.

2. Not only the collection of old appliances can be improved. An efficient organization of the complete recycling chain is also very important. This means that only high quality processes with which many elements can be efficiently recovered should be used. At the same time the high social and environmental standards and regulations must be met. Although the recycling chain now includes efficient recovery processes for nearly all materials, both at the mechanical stage of sorting and preprocessing as well as in the metallurgical step, in many cases

they are not implemented. The economics of the process are important and must be understood, not so much in comparison to the primary exploitation of the raw materials, but more especially to the use of cheaper recycling processes that are also less efficient. This is often possible because hazardous emissions are not controlled, hence the costs for observing the environmental and social regulations do not come into effect, and therefore are not internalized into the cost calculation of the recovery process. This problem pertains to the 25–30% of electronic scrap and end-of-life appliances that are estimated to be illegally exported from Europe to regions without appropriate recycling infrastructure and standards.

3. The design for recycling (recycling-friendly design of appliances) and also the design for disassembly (specific components are easy to extract from the appliance) can be improved in many products without compromising the performance of the appliance. The extreme complexity of the "urban mine" can be partially reduced with such forward-looking approaches. For example, if magnets, batteries and electronic components are easier to access in automobiles or electrical appliances, then they are easier to dismantle and extract before shredding and fed into special recovery processes (Fig. 3.29), as there are clear physical limitations on the recovery of trace elements after the shredding process.

4. The recovery rates for some metals are still today unsatisfactory, especially for the rare-earth elements or those metals, for which there is only a limited specialized metallurgical infrastructure. These are the metals in the two outer rings on the "metal wheel", such as the important electronic metals gallium, germanium or indium. The recovery rates must be significantly improved, also with respect to the supply of these raw materials to the German markets. The BMBF r^4 research program "Raw Materials of Strategic Economic Importance for High-Tech Made in Germany" [52] contributes to this issue.

3.4.6   The Influence of Substitution and Increased Material Efficiency on the Supply of Natural Resources

In addition to the exploitation of natural resources from primary deposits in the geosphere and from secondary deposits in the technosphere, substitution is the third pillar for ensuring the supply. The substitution processes are mainly driven by the feedback control cycle of raw material supply that is described in Sect. 2.4. Today there are generally five categories of substitution measures that all have the objective of replacing completely, or partially, a specific natural resource [129]:

1. *Material substitution*: A material or an element is replaced by another
2. *Technological substitution*: The use of a material is reduced by technological advances and improvements in the manufacturing process, but the functionality is retained.
3. *Functional substitution*: A product is replaced by another product with the same functionality.

4. *Quality substitution*: High-quality products are replaced by economizing on some materials in the manufacturing of lower quality products.
5. *Non-material substitution*: The use of materials is reduced by increasing the non-material factors such as work and energy.

The measures 1–3 are of interest for the energy systems of the future, but measures 4 and 5 are not of interest.

An example for material substitution (1) is replacing copper with aluminum for the transmission of electricity. The miniaturization of tantalum-bearing capacitors in the electronics industry, such as for cell phones, is an example of technological substitution. The replacement of synchronized motors, in which the magnets contain rare-earth elements, is an example of functional substitution by the so-called squirrel-cage asynchronous motors or ferrite motors, which do not contain rare-earth elements.

BOX XI: Metal Wheel

The metal wheel is a presentation in which the various chemical elements are classified according to the main metallurgical processes (Fig. 3.30). The individual elements are subdivided into rings in the metal wheel that reflect, on the one hand, the market value and the concentration of the elements in ores and, on the other hand, the chemical characteristics of the element, such as the oxidation capability or, as for copper, the ability to act as a collector for other metals, and the thermodynamic conditions in the relevant metallurgical processes (loss as a gas phase or in the slag).

The innermost ring of the wheel consists of metals for which the traditional metallurgical processes were primarily developed. This includes the oxide ores of aluminum (Al), chrome (Cr), iron (Fe), magnesium (Mg), manganese (Mn), titanium (Ti) and tin (Sn), the sulfide ores of lead (Pb), copper (Cu) and zinc (Zn), and the both oxide and sulfide ores of nickel (Ni). The next ring (I) includes those metals that coexist with the primary metals in their deposits, for which there is usually a connected and significant production infrastructure. These metals have a high economic value and are sometimes required in high-technology products. These metals include the high-value precious metals gold (Au) and silver (Ag), which are recovered in the smelting of lead (Pb), copper (Cu), nickel (Ni) and zinc (Zn). The next ring (II) consists of those metals that occur as by-products, for which there is a limited or even no specific metallurgical infrastructure. Many of these metals are valuable as high-technology and electronic metals. The outer ring (III) includes those by-product metals for which there is as yet no infrastructure, and they are usually lost in the slags or the emissions from the smelting of other metals. Antimony (Sb) for example is usually lost to the slags of copper or nickel smelting. The metallurgical process for this metal is very cost-intensive because the recovery process is at the end of the production chain, and requires another technology.

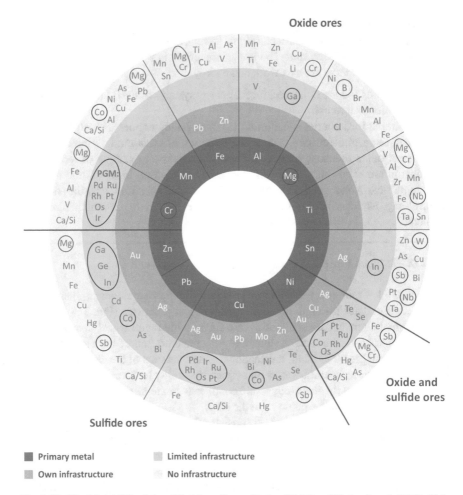

Fig. 3.30 The Metal Wheel (modified from Reuter/Verhoef [124] and Verhoef et al. [125]). This presentation was initially based on the smelting and refining processes for primary ores, but it can also be used for secondary resources [126]. The raw materials identified by the EU Commission as critical in their lists from 2010 and 2014 (EC: list for the EU with 14 critical raw materials [127]; EC: list for the EU with 20 critical raw materials [128]) are color-coded for orientation: the red-circled elements were identified as critical in 2010; the blue-circled elements were added in 2014; and the green-circled tantalum is the only metal to have been dropped from the lists. *Source* Reuter and Verhoef [124] and Verhoef et al. [125], with permission of the authors

The different metallurgical processes have been developed over centuries for ores with particular compositions that occur as natural deposits in the geosphere. However, there are no recovery processes available (due to thermodynamic constraints) for several valuable by-product metals from certain types of mineral deposit. For example, it is still not possible to recover tantalum directly from tin ores. The tantalum reports directly to the tin slags and must be recovered in a separate process, so long as it is still economic. This emphasizes that the new and "random" compositions of elements in secondary materials in many cases represent much greater challenges for metallurgy. A 100% recycling of all elements from a complex mixture of materials is not possible. Depending on the selected processing route, there will always be some losses of some substances because the output cannot be optimized for all metals.

Because every element has its own specifications, there are always uses in which an element substitution is only the second-best solution. Material flow analyses [130] demonstrate that there is no element which can be perfectly substituted in all its uses by another element (Fig. 3.31). A recent study [131] systematically analyzes and categorizes the possibilities for product and usage relevant substitutions for the list of critical raw materials [127] that was prepared by the European Commission in 2010. The possibilities for substitution are generally only possible for specific products, and it is therefore very difficult to make general statements on the possibilities for the substitution of one raw material. In these studies, the results of which are the basis for the summary in Fig. 3.31, the possibilities for substitutions of individual resources are weighted and classified according to their various uses. The so derived substitution success for one element only provides an indication of its potential, and therefore also for the energy systems of the future, but cannot make any statement on individual cases. One raw material that according to this global view has excellent substitution characteristics, might for one manufacturer be an irreplaceable raw material for which there are no alternatives for its specific use, and therefore it is a critical raw material. The lower the sum of the substitution possibilities for the raw material in a product, then the risks for the manufacturer of this specific product are greater during supply shortages.

Figure 3.31 presents a static analysis and does not provide any indication of the innovation potential. It is quite possible that a shortage of raw material and increased prices will, according to the feedback control cycle of raw material supply, initiate an intensive investigation for substitution possibilities and the development of appropriate technologies. The solutions for the scarcity of cobalt, which resulted from the Shaba crisis in 1978 [130], or more recently that of rhenium [132], are examples of this reaction. In the case of cobalt, new alloys with permanent magnetic characteristics [133, p. 33] were discovered, and alternative alloy compositions were identified for rhenium.

If a one-for-one substitution is not possible, but only a functional substitution, then it takes some time before an alternative product is developed, and it will be several years before its industrial application.

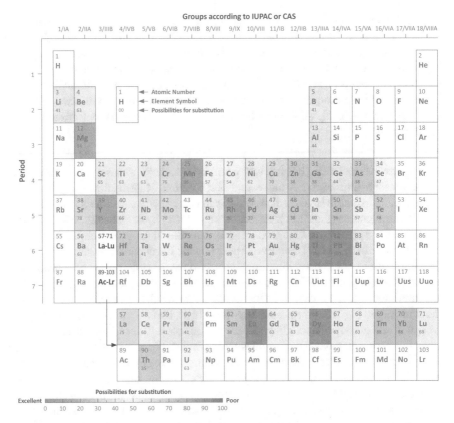

Fig. 3.31   Average efficiency of the direct Substitution of Elements (Graedel et al. [130], Fig. 5). The substitution success is presented as colors for the range from 0 to 100. 0 indicates that there are replacement materials for the main applications. 100 means that there is no adequate possibility for substitution of any application. Ultimately, only the testing of each individual case is decisive. *Source* Graedel et al. [130], with permission of PNAS

The Supply of Secondary Resources and Development of Technologies for Recycling

There is less dependency on the imports of primary resources if the demand for raw materials can be met, as much as possible, by recycling domestic end-of-life materials. However, there is still significant potential that is not used.

As compared to the exploitation of primary resources, the recovery of secondary raw materials has different requirements because of the material compositions and various construction techniques in products. A "100-% recycling" is not possible because the recovery of metals from complex material compositions is constrained by physical limits. This, together with the residence times of materials in the economic cycle, are the reasons that it is not possible to completely meet the raw material requirements from secondary source materials.

The losses of raw materials during recycling can be minimized by a product design appropriate for recycling, by establishing material cycles, globally harmonized quality and environmental standards for the recycling processes and by facilitation of shipments to state- of- the- art recycling plants. A circular economy must be established so that the metals introduced into the technosphere can be, as much as is possible, be channeled into the appropriate recycling process-chain. Only so can the maxim "use metals, instead of consuming metals" be practically implemented by the society.

The supply of secondary resources as well as a sufficient quality of scrap represent a challenge for the German and European smelters: the rapid life cycle of many (electronic) appliances requires them to adapt to the changing compositions of the scrap material by establishing flexible processes. The overall recycling efficiency depends on the efficiency of the separate processing steps. The efficiency of the first step, the collection of material, is often low because there is no incentive for the consumers to deliver their end-of-life appliances to recycling.

The geo- and technospheres have sufficient potential to meet an increasing demand for resources until 2050. This is conditional on a functioning market economy, efficient and sustainable infrastructures as well as research and development.

The Limits to Substitution

The substitution of materials that are considered to be high risk with lower risk materials is currently being discussed in the EU. This issue has been the particular focus of the latest study [42, p. 142ff] from the Institute for Energy and Transport (JRC-IET) within the Joint Research Center (JRC) of the European Commission. The supply risk is obviously an abstract threat, and therefore every company must decide for themselves how they can successfully compete with the best products in the global market by minimizing the risks and maximizing their chances for business success. Therefore a company must evaluate the raw material risk in a larger framework. In some cases, it may not be advantageous to replace critical with less critical raw materials. In its study *"Critical mineral resources and material flow during the transition of the German energy supply system (KRESSE)* [134, p. 186]", the Wuppertal Institute for Climate, Environment and Energy concludes that "the application of possible critical mineral resources can be advantageous for the specific resource efficiency of technologies if, as a result, systems are thereby overall more material and energy efficient".

BOX XII: Political Incentives for Material Efficiency

The Federal Ministry for Economic Affairs and Energy (BMWi) awards the "German Prize for Resource Efficiency" to promote more strongly the significance of natural resource and material efficiency and their effects on the economy. The German Mineral Resources Agency has been commissioned by the BMWi since 2011 with the organization and justification. The prize commends excellence for outstanding industrial examples of resource and raw materials efficiency in products, processes and services as well as the results of applied research along the whole value chain from resource exploitation and processing to recycling.

The issue of resource efficiency as a contribution to the sustainable exploitation and use of resources is of central importance to both the Federal Government's resource strategy of October 2010 [135] and the EU communication of February 2011 [136]. In addition to an improvement in resource efficiency, a further political objective of the Federal Government is to double resource productivity in Germany by 2020 (based on 1994 as the reference year with value 100). In 2013 this value had increased to 147.8 [137, p. 41] (Federal Agency for Statistics). The resource productivity is the ratio of the gross domestic product and the total consumption of resources and materials. This means that a reduction in the quantity of required resources for the same added value, or an increase of the added value from the same quantity of resources, leads to an increase in the resource productivity. The indicator is used, among others, in the Federal Government's[29] national sustainability strategy "Perspectives for Germany".

The international Factor-X movement should also be mentioned in this respect because it also has the objective to improve the total resource efficiency in our society. The international Factor-10 Institute [138] is the best-known example of the Factor-X movement, and has the objective to increase the resource efficiency in society by a factor of ten. The increase of the material efficiency is an important objective with respect to sustainability, and has a positive effect on the overall resource situation. However, the development of future energy systems is initially expected to require an overall increase in the resource requirements, and efficiency measures could primarily lower the overall costs.

One of the substitution measures, namely technological substitution, also relates directly to an increase in material efficiency because it requires less resources. The most effective incentive is, as always, saving material costs. In the case of transitioning energy systems to renewable technologies it is practically impossible to realize this requirement during the early stages. The manufacturing of facilities for gener-

---

[29] Federal Government 2002 [139].

ating renewable energy is, relative to the generated unit of energy, in many cases initially more resource intensive as for facilities generating conventional energy. This initially counteracts the improvements in resource productivity [17, 18, 19]. The improved resource efficiency of renewable energy becomes apparent over the total life-time cycle, from initial investment to the end-of-life, because no fossil energy resources are used for the generation of renewable energy. This exemplifies how the initial investment in more resources ultimately results in improved resource efficiency [60].

Substitution and Increasing Material Efficiency

Industry can react more flexibly to raw materials shortages by searching the entire added-value chain for ways to substitute the resource most liable to shortage risks or to use these raw materials more efficiently. In addition to avoiding the use of these risk-related raw materials and improving the material efficiency, the customers can also be persuaded in advance to accept another material composition in the products in certain circumstances ("product clearance"). An intensification of research into materials is fundamental for successful substitution and avoidance strategies. This will ultimately result in a strengthening of the perception of high-technology in Germany, particularly for the small and medium-size enterprises.

References

(Note: All Web links listed were active as of the access date but may no longer be available.)

1. Crowson, P.: "Solving the Minerals Equation? Demand, Prices and Supply". In: *LE STUDIUM® Conference: Life and Innovation Cycles in the Field of Raw Materials Supply and Demand – a Transdisciplinary Approach*, Orléans, France, 19. – 20. April 2012.
2. Bundesanstalt für Geowissenschaften und Rohstoffe: *Deutschland – Rohstoffsituation 2013*, Hannover 2014.
3. Commodity Research Bureau: *Market Data*, 2013. URL: http://www.crbtrader.com/data.asp?page=chart&sym=BTY00&name=BLSMetals&domain=crb&display_ice=1&studies=-Volume;&cancelstudy=&a=M [accessed: 14.10.2013].
4. SNL Metals & Mining (formerly: Metals Economics Group – MEG): *World Wide Exploration Trends 2013*, Halifax 2013. URL: http://www.metalseconomics.com/sites/default/files/uploads/PDFs/meg_wetbrochure2013.pdf [accessed: 14.10.2013].
5. Deutsche Rohstoffagentur in der Bundesanstalt für Geowissenschaften und Rohstoffe: *Statische "Reichweite" und globale Bergwerksförderung am Beispiel einzelner Rohstoffe* (Data compilation from the data banks of the Bundesanstalt für Geowissenschaften und Rohstoffe and of the United States Geological Survey), 2014.
6. Bundesanstalt für Geowissenschaften und Rohstoffe: *BGR-Datenbank*, Hannover: Bundesanstalt für Geowissenschaften und Rohstoffe 2014.
7. Dorner, U./Schmidt, M./Liedtke, M./Buchholz, P.: *Frühwarnindikatoren und Rohstoffrisikobewertung – Methodischer Überblick am Beispiel Antimon* (Commodity Top News

43), Deutsche Rohstoffagentur in der Bundesanstalt für Geowissenschaften und Rohstoffe 2014.
8. Frenzel, M./Ketris, M.P./Gutzmer, J.: "On the geological Availability of Germanium". In: *Mineralium Deposita*, 49: 4, 2014, p. 471–487.
9. US Geological Survey: *Mineral Commodity Summaries 2015*, Washington DC 2014. URL: http://minerals.usgs.gov/minerals/pubs/mcs/2015/mcs2015.pdf [accessed: 07.08.2015].
10. US Geological Survey: *Mineral Commodity Summaries 2001*, Washington DC 2001. URL: http://minerals.usgs.gov/minerals/pubs/mcs/2001/mcs2001.pdf [accessed 28.10.2014].
11. Hubbert, M.K.: "Nuclear Energy and Fossil Fuels". In: *Proceedings of American Petroleum Institute Drilling & Production Practice* (spring meeting), San Antonio, Texas, 1956, p. 7–25.
12. Wellmer, F.-W.: "Wie lange reichen unsere Rohstoffvorräte? – Was sind Reserven und Ressourcen?". In: *umweltforum (uwf)*, 22, 2014, p. 125–132.
13. Ericsson, M./Söderholm, P.: *Mineral Depletion and Peak Production* (POLINARES (EU Policy in Natural Resources) working paper n.7, D1.1 – Partial Report of Working package 1 "Framework for Understanding the Sources of Conflict and Tension"), 2010. URL: http://www.polinares.eu/docs/d1-1/polinares_wp1_peak_debates_minerals.pdf [accessed: 28.10.2014].
14. Wellmer, F.-W./Dalheimer, M./Wagner, M.: *Economic Evaluations in Exploration*, Berlin, Heidelberg, New York: Springer Verlag 2008.
15. McLean, H.L./Duchin, F./Hagelüken, C./Halada, K./Kessler, S.E./Moriguchi, Y./Mueller, D./Norgate, T.E./Reuter, M.E./van der Voet, E.: "Stocks, Flows and Prospects of Mineral Resources". In: Graedel, T.E./van der Voet, E. (Eds.): *Linkages of Sustainability*, Strüngmann Forum Report, Cambridge: MIT Press, 2010, pp. 199–218.
16. Rosas, J./Schuffeneger, C./Cornejo, C.: "Update of Chilean Mining Projects and technological Trends", In: *44th Annual Canadian Mineral Processors Operators conference*, Ottawa, 17–19. January 2012.
17. Wellmer, F.-W.: "L'enseignement des géosciences minières en Europe". In: *Géosciences*, 15, 2012, pp. 100–101.
18. Vidal, O./Goffé, B./Arndt, N.: "Metals for a low-carbon Society" (Supplementary Information). In: *Nature Geoscience*, 6, 2013, pp. 894–896.
19. Hertwich, E.G./Gibojn, T./ Bouman, E. A./Arvesen, A./Suh, S./Heath, G. A./Bergesen, J. D./Ramirez, A./Vega, M. I./Shi, L.: "Integrated life-cycle Assessment of Electricity-Supply Scenarios confirms global environmental Benefit of low-carbon Technologies". In: *Proceedings of the National Academy of Sciences of the USA*, 2014. URL: http://www.pnas.org/content/suppl/2014/10/02/1312753111.DCSupplemental [accessed: 28.10.2014].
20. Skinner, B. J.: "Earth Resources". In: *Proceedings of the National Academy of Sciences of the USA*, 76: 9, 1979, pp. 4212–4217.
21. Verein Deutscher Ingenieure – Fachbereich Ressourcenmanagement (VDI): *VDI 4600: Kumulierter Energieaufwand (KEA) – Begriffe, Berechnungsmethoden*, Berlin: Beuth Verlag 2012.
22. Teuber, J./Hofmann, M./Kosinowski, M./Sattler, H./Schumacher, K.: "Der kumulierte Energieaufwand für die Erdölgewinnung am Beispiel ausgewählter Felder des Gifhorner Troges". In: Proceedings *DGMK- Spring meeting Celle*, 1999, S. 31–40.
23. Hall, C.A.S./Balogh, S./Murphy, D.J.R.: "What is the minimum EROI that a sustainable Society must have?". In: *Energies*, 2, 2009, pp. 25–47.
24. Buchholz, P.: "Entwicklung von Frühwarnindikatoren für die Rohstoffrisikoanalyse". In: *BGR- Rohstoffkonferenz 2013*, Hannover, 4. November 2013.
25. Bräuninger, M./Leschus. L./Rossen, A.: "Ursachen von Preispeaks, -einbrüchen und -trends bei mineralischen Rohstoffen". In: *DERA Rohstoffinformationen Nr. 17*, Commissioned study by Hamburgische WeltWirtschaftsInstitut (HWWI) for the Deutsche Rohstoffagentur in der Bundesanstalt für Geowissenschaften und Rohstoffe, Berlin 2013.
26. Buchholz, P.: *Marktüberblick, Preisentwicklung und Erhöhung der Versorgungssicherheit*, Vortrag beim Bundesverband Materialwirtschaft, Einkauf und Logistik, Chemie Einkauf, Frankfurt am Main, 20. November 2013.

27. Stürmer, M.: *150 Years of Boom and Bust: What Drives Mineral Commodity Prices*, MPRA Working Paper, 2013.
28. Stürmer, M.: "Industrialization and the Demand for Mineral Commodities". In: *Bonn Econ Discussion Papers*, 13/2013, 2013.
29. Bleischwitz, R.: "International Economics of Resource Productivity – Relevance, Measurement, empirical Trends, Innovation, Resource Policies". In: *Int. Economics and Economic Policy*, 7: 2–3, 2010, p. 227–244.
30. Wellmer, F.-W./Dalheimer, M.: "The Feedback Control Cycle as Regulator of past and future Mineral Supply". In: *Mineralium Deposita*, 47: 7, 2012, pp. 713–729.
31. National Research Council of the National Academies: *Minerals, Critical Minerals, and the U.S. Economy*, Washington, D.C.: The National Academies Press 2008.
32. Malenbaum, W.: *World Demand for Raw Materials in 1985 and 2000*, New York: McGraw-Hill 1978.
33. Stürmer, M./von Hagen, J.: *Einfluss des Wirtschaftswachstums aufstrebender Industrienationen auf die Märkte mineralischer Rohstoffe – Entwicklung von Bewertungsgrundlagen am Beispiel ausgewählter sich schnell industrialisierender Staaten der Vergangenheit und der BRIC-Staaten*, Expert opinion for the Bundesanstalt für Geowissenschaften und Rohstoffe. Institut für Internationale Wirtschaftspolitik (DERA-Rohstoffinformation 11), Berlin: Deutsche Rohstoffagentur in der Bundesanstalt für Geowissenschaften und Rohstoffe 2012. URL: http://www.deutsche-rohstoffagentur.de/DE/Gemeinsames/Produkte/Download s/DERA_Rohstoffinformationen/rohstoffinformationen-11.pdf?__blob=publicationFile&v= 3 [accessed: 11.08.2014].
34. Kravis, I. B./Heston, A. W./Summers, R.: "Real GDP per Capita for more than one hundred Countries". In: *The Economic Journal*, 88: 350, 1978, p. 215–242.
35. Statistisches Bundesamt: *Statistisches Bundesamt in Cooperation with eurostat*, 2013. URL: https://www.destatis.de/Europa/DE/Thema/UnternehmenProduktion/Industrie.html [accessed: 22.10.2014].
36. Bundesministerium für Bildung und Forschung: *Ideen. Innovation. Wachstum - Hightech-Strategie 2020 für Deutschland*, Bonn: BMBF 2010. URL: http://www.bmbf.de/pub/hts_20 20.pdf [accessed: 22.10.2014].
37. Bundesministerium für Bildung und Forschung: *Die Neue Hightech-Strategie – Innovationen für Deutschland,* Berlin: BMBF 2014. URL: http://www.bmbf.de/pub_hts/HTS_Broschure_ Web.pdf [accessed: 16.09.2015].
38. Bayerische Motoren Werke AG: personal communication, 11.07.2014.
39. Pischetsrieder, B.: personal communication, acatech Workshop Rohstoffe – Werkstoffe – Neue Technologien, Berlin, 26.10.2011.
40. National Research Council of the National Academies: *Minerals, critical minerals, and the U.S. economy. Prepublication Version*, Washington D.C.: The National Academies Press 2007. URL: http://www.nma.org/pdf/101606_nrc_study.pdf [accessed: 27.10.2014].
41. Hagelüken, C.: *Bedeutung des Recyclings für die Verfügbarkeit von strategischen Metallen* (Paper WING-Conference, Session 3: Strategische Metalle, Rohstoffe, Substitution, Chemie, 05.10.2011, Berlin), 2011.
42. Moss, R. L./Tzimas, E./Willis, P./Arendorf, J./Tercero Espinoza, L.: *Critical Metals in the Path towards Decarbonisation of the EU Energy Sector – Assessing Rare Metals as Supply-Chain Bottlenecks in Low-Carbon Energy Technologies* (Scientific and Policy Reports), Petten: European Commission, Joint Research Centre, Institute for Energy and Transport 2013. URL: https://setis.ec.europa.eu/newsroom-items-folder/new-jrc-report-critical-metals-energ y-sector [accessed: 02.05.2014].
43. Melcher, F./Buchholz, P.: "Current and future Germanium Availability from primary Resources". In: *Minor Metals Conference*, Köln, 24.04.2012.
44. Bundesanstalt für Geowissenschaften und Rohstoffe: *BGR-Datenbank*, Hannover 2015. Metal Bulletin: Price Book (Online-Data bank), London 2015. http://www.375metalbulletin.com/ [accessed: 20.09.2015].

45. Metal Bulletin: *Price Book* (Online-Data bank), London 2015. URL: http://www.metalbulle
    tin.com/ [accessed: 20.09.2015].
46. Displaysearch: DisplaySearch Reports Q4'07 Worldwide LCD TV Shipments Surpass
    CRTs for First Time; TV Revenues Reach a Record High, Up 10% to $33B, Display-
    Search/Information Handling Services (IHS) 2008. URL: http://www.displaysearch.com/c
    ps/rde/xchg/displaysearch/hs.xsl/6138.asp [accessed: 25.09.2014].
47. Lachmund, H.: *Alternative Analysis and alloying Strategies in heavy plating Processing as
    a Result of alloying Agents Price Trends,* Paper: Life and Innovation Cycles in the Field
    of Raw Materials Supply and Demand – a Transdisciplinary Approach, Orléans, France,
    18.–19.04.2012.
48. Angerer, G./Erdmann, L./Marscheider-Weidemann, F./Lullmann, A./Scharp, M./Handke,
    V./Marwede, M.: *Raw Materials for emerging Technologies,* Stuttgart: Fraunhofer IRB
    Verlag 2009. URL: http://www.isi.fraunhofer.de/isi-en/service/presseinfos/2009/pri09-02.ph
    p [accessed: 27.10.2014].
49. Angerer, G./Erdmann, L./Marscheider-Weidemann, F./Lullmann, A./Scharp, M./Handke,
    V./Marwede, M.: *Rohstoffe für Zukunftstechnologien: Einfluss des branchenspezifischen
    Rohstoffbedarfs in rohstoffintensiven Zukunftstechnologien auf die zukünftige Rohstoffnach-
    frage,* Stuttgart: Fraunhofer IRB Verlag 2009. URL: http://www.isi.fraunhofer.de/isi-en/serv
    ice/presseinfos/2009/pri09-02.php [accessed: 27.10.2014].
50. Marscheider-Weidemann, F., Langkau, S., Hummen, T., Erdmann, L., Tercero-Espinoza, L.,
    Angerer, G., Marwede, M., Benecke, S.: Rohstoffe für Zukunftstechnologien 2016: *DERA
    Rohstoffinformationen Nr. 28,* Berlin: Deutsche Rohstoffagentur in der Bundesanstalt für
    Geowissenschaften und Rohstoffe 2016.
51. Bundesministerium für Bildung und Forschung: *Forschung für nachhaltige Entwick-
    lungen (FONA) – Rahmenprogramm des BMBF,* Bonn, Berlin: BMBF 2009,
    p. 59. URL: http://www.fona.de/mediathek/pdf/forschung_nachhaltige_entwicklungen_neu.
    pdf [accessed: 29.10.2014].
52. Bundesministerium für Bildung und Forschung: *Werkstoffinnovationen für Industrie und
    Gesellschaft (WING) – Rahmenprogramm des BMBF,* Bonn: BMBF 2003. URL: http://www.
    bmbf.de/pub/rahmenprogramm_wing.pdf [accessed: 29.10.2014].
53. Bundesministerium für Bildung und Forschung: *Ressourceneffizienz potenzieren. Broschüre
    zum Förderschwerpunkt "Innovative Technologien für Ressourceneffizienz – rohstoffinten-
    sive Produktionsprozesse" (r^2),* Karlsruhe: Fraunhofer-Institut für System- und Innovations-
    forschung ISI 2010. URL: http://www.r-zwei-innovation.de/_media/r2_broschuere_web.pdf
    [accessed: 29.10.2014].
54. Bundesministerium für Bildung und Forschung: *Wirtschaftsstrategische Rohstoffe für den
    Hightech-Standort Deutschland (r^4),* Bonn: BMBF 2012.
55. Bundesverband der Deutschen Industrie e.V.: *Anforderungen an eine ganzheitliche und nach-
    haltige Rohstoffpolitik – BDI Grundsatzpapier zur Rohstoffpolitik im 21. Jahrhundert,* Berlin:
    BDI 2015.
56. Schmidt, M.: "Rohstoffrisikobewertung – Platingruppenelemente Platin, Palladium, Rhodi-
    um". In: *DERA Rohstoffinformationen 26,* Berlin: Deutsche Rohstoffagentur in der Bun-
    desanstalt für Geowissenschaften und Rohstoffe 2015. URL: http://www.deutsche-rohs
    toffagentur.de/DERA/DE/Publikationen/Schriftenreihe/schriftenreihe_node.html [accessed:
    12.10.2015].
57. Christlich Demokratische Union Deutschlands/Christlich-Soziale Union in Bay-
    ern/Sozialdemokratische Partei Deutschland: *Deutschlands Zukunft gestalten* (Coalition
    agreement between CDU, CSU und SPD, 18. legislative period. 16.12.2013), Berlin:
    Coalition parties of the Federal Government 2013.
58. Buchholz, P./Huy, D./Liedtke, M./Schmidt, M.: *DERA-Rohstoffliste 2014, Angebotskonzen-
    tration bei mineralischen Rohstoffen und Zwischenprodukten – Potenzielle Preis- und Lie-
    ferrisiken* (DERA Rohstoffinformationen Nr. 24), Berlin, Deutsche Rohstoffagentur in der
    Bundesanstalt für Geowissenschaften und Rohstoffe 2015. http://www.deutsche-rohstoffag
    entur.de/DERA/DE/Publikationen/Schriftenreihe/schriftenreihe_node.html [Stand: accessed
    04.2015].

59. Bundestags-Enquete-Kommission: *Schutz des Menschen und der Umwelt – Ziele und Rahmenbedingungen einer nachhaltig zukunftsverträglichen Entwicklung*, Deutscher Bundestag: Abschlussbericht der Enquete-Kommission, Bundestagsdrucksache 13/ 11200 of 26.06.98, Berlin 1998.

60. Wagner, M./Wellmer, F.-W.: "A Hierarchy of Natural Resources with Respect to Sustainable Development – A Basis for a Natural Resources Efficiency Indicator". In: Richards, J. P.: *Mining, society and a Sustainable World*, Berlin, Heidelberg: Springer Verlag 2009, pp. 91–121.

61. Liedtke, M./Elsner, H: "Seltene Erden". In: *Commodity Top News Nr. 31*, Hannover: Bundesanstalt für Geowissenschaften und Rohstoffe (BGR) 2009.

62. Fischer, H. (Brose Fahrzeugteile GmbH & Co. KG): personal communication, 01.07.2014.

63. Liedtke, M./Schmidt, M.: "Rohstoffrisikobewertung – Wolfram", In: *DERA Rohstoffinformationen Nr. 19*, Berlin: Deutsche Rohstoffagentur in der Bundesanstalt für Geowissenschaften und Rohstoffe 2014.

64. World Trade Organization: *China – Measures related to the Exportation of Rare Earths, Tungsten, and Molybdenum*, Reports of the Appellate Body: WT/DS431/AB/R, WT/DS432/AB/R, WT/DS433/AB/R, Geneva 2014. URL: https://www.wto.org/english/tratop_e/dispu_e/cases_e/ds431_e.htm [accessed: 22.01.2016].

65. World Trade Organization: *China – Measures related to the Exportation of various Raw Materials*, Reports of the Appellate Body: WT/DS394/AB/R, WT/DS395/AB/R, WT/DS398/AB/R), Geneva 2012. URL: https://www.wto.org/english/tratop_e/dispu_e/cases_e/ds394_e.htm [accessed: 22.01.2016].

66. US Department of Energy: *Critical Materials Strategy*, Washington DC 2011. URL: http://energy.gov/sites/prod/files/DOE_CMS2011_FINAL_Full.pdf [accessed 27.06.14].

67. Firebreak 2013. URL: http://www.borax.com/product/firebrake-zb.aspx [accessed 14.3.2014].

68. Siemens: "Materials for Magnets in Wind Turbines", In: Metal Events – 10th International Rare Earths Conference, Singapore, 16.10.2014.

69. Goskowski, F.: *Three Essays and three Revolutions*, Durham CT: Strategic Book Group 2011.

70. Moody, R.: *Schmutzige Geschäfte – Deutsche Investitionen im Bergbausektor*, Berlin: Heinrich-Böll-Stiftung 2015. URL: http://www.boell.de/sites/default/files/2014-03-deutsch e-investitionen_bergbausektor.pdf [accessed: 13.03.2015].

71. Deutsche Gesetzliche Unfallversicherung: *Geschäftsergebnisse*, 2000–2012.

72. Deutsche Gesetzliche Unfallversicherung: *Meldepflichtige Arbeitsunfälle je 1 Mio. geleisteter Arbeitsstunden*, 2014. URL: http://www.dguv.de/de/Zahlen-und-Fakten/Arbeits-und-Wegeu nfallgeschehen/Meldepflichtige-Arbeitsunf%C3%A4lle-je-1-Mio.-geleisteter-Arbeitsstunde n/index.jsp [accessed: 11.04.2014].

73. Nelles, M.: "Wirtschaftliche und technologische Herausforderungen im modernen Erzbergbau – Strategien und Lösungsansätze eines mittelständischen Bergbauunternehmens". In: *3. Sächsischer Rohstofftag*, Freiberg, Sachsen, 02.04. 2008.

74. Werner, W.: "Schätze unter dem Boden: Was wissen wir über die tiefliegenden Rohstoffe in Baden-Württemberg". In: *Berichte der Naturforschenden Gesellschaft zu Freiburg i. Br., 102*, 2012, pp. 37–92. URL: http://www.lgrb-bw.de/download_pool/naturf_ges_band_102_s37-9 2_2012.pdf [accessed: 13.11.2014].

75. Wellmer, F.-W.: "Sustainable Development and Mineral Resources". In: *Géosciences*, 15, 2012, pp. 8–14.

76. Bundesverfassungsgericht: *1 BvR 3139/08, 1 BvR 3386/08, Urteil des Ersten Senats vom 17.12.2013, Absatz-Nr. (1-333)*, Karlsruhe 2013. URL: http://www.bverfg.de/entscheidunge n/rs20131217_1bvr313908.html [accessed: 12.08.2014].

77. Gauthier, M.: *École des Mines*, Paper Ressources minérales: la vision du mineur, 01.02.2012, Paris 2012.

78. Rankin, W. J.: "Towards zero Waste – Re-Evaluating the traditional Production Cycle". In: *The Australasian Institute of Mining and Metallurgy Bulletin* (AusIMM), June 2015. URL: https://www.ausimmbulletin.com/feature/towards-zero-waste/ [accessed: 14.10.2015].

79. Siefert, M.: *Erfahrungen mit innovativen Abbaukonzepten im alpinen untertägigen Bergbau*, Paper at the Honorary Colloquium "Bergbau gestern – heute – morgen" on the occasion of the 90th birthday of Prof. G. B. L. Fettweis und the 75th birthday of Prof. H. Wagner, Leoben, 20.11.2014.

80. Misereor: *Menschenrechtliche Probleme im peruanischen Rohstoffsektor und die deutsche Mitverantwortung*, Aachen 2013. URL: http://www.misereor.de/themen/wirtschaft-fuer-die-armen/rohstoffe/menschenrechts-verletzungen-im-bergbau-in-peru.html [accessed 15.02.2015].

81. Hamm, B./Schax, A./Scheper, C.: *Human rights Impact Assessment of the Tampakan Copper-Gold Project*, Mindanao, Philippines 2013. URL: http://www.misereor.de/themen/wirtschaf t-fuer-die-armen/rohstoffe.html [accessed: 13.04.2014].

82. commdev.org: *Codes, Standards and Guidluines*, CommDev is an initiative of the International Finance Corporation (IFC), 2015. URL: http://commdev.org/codes-standards-and-guidelines [accessed: 18.02.2015].

83. International Council on Mining & Metals: *10 Principles*, International Council on Mining & Metals, 2015. URL: http://www.icmm.com/our-work/sustainable-development-framewor k/10-principles [accessed: 18.02.2015].

84. Global Reporting Initiative (GRI): *Guidance for Reporting in the Mining and Metals Sector*, 2014. URL: https://www.globalreporting.org/reporting/sector-guidance/sector-guidance/min ing-and-metals/Pages/default.aspx [accessed: 2.3.2014].

85. Globaldialogue: *Intergovernmental Forum of Mining, Minerals, Metals and sustain-able Development*, 2015. URL: http://www.globaldialogue.info/intro_e.htm [accessed: 24.08.2015].

86. Prno, J.: "An Analysis of Factors leading to the Establishment of a Social Licence to Operate in the Mining Industry". In: *Resources Policy*, 38, 2013, pp. 577–590.

87. EY: *Business Risks facing Mining and Metals 2015-2016. Industries – Mining and Met-als*, 2015. URL: http://www.ey.com/GL/en/Industries/Mining—Metals/Business-risks-in-mi ning-and-metals [accessed: 15.12.2015].

88. Steinbach, V.: "Responsibility in the Mineral Resources Sector – Engine for the global Development and social Progress", In: *Internationale Rohstoffkonferenz "Verantwortung übernehmen – Nachhaltigkeit in der Rohstoffwirtschaft fördern"*, Berlin, 10. – 11.11.2015. URL: http://www.bgr.bund.de/EN/Themen/Min_rohstoffe/Veranstaltungen/Rohstoffkonfere nz2015/Raw_Material_Conference_2015_Liste.html?nn=7050122 [accessed: 04.01.2016].

89. Franken, G./Vasters, J./Dorner, U./Melcher, F./Sitnikova, M./Goldmann, S.: "Certified Trad-ing Chains in Mineral Production: A Way to improve Responsibility in Mining". In: R. Sinding-Larsen, F.-W. Wellmer (Eds.): *Non-Renewable Resource Issues – Geoscientific and Societal Challenges*, Dordrecht, Heidelberg: Spinger Verlag 2012, pp. 213–227.

90. The Equator Principles Association: *Environmental and social Risk Management for Projects*, 2014. URL: http://www.equator-principles.com/ [accessed: 02.03.2014].

91. Transparency Deutschland: *Corruption Perceptions Index 2013*, Berlin 2014. URL: http:// www.transparency.de/Tabellarisches-Ranking.2400.0.html [accessed: 13.11.2014].

92. Bundesministerium für wirtschaftliche Zusammenarbeit und Entwicklung: *Gute Regierungs-führung und nachhaltige Entwicklung lassen sich nicht trennen*, BMZ 2015. URL: http:// www.bmz.de/de/was_wir_machen/themen/goodgovernance/ [accessed: 19.02.2015].

93. G7: *The Brussels G7 Summit Declaration*, Brussels 2014. URL: http://www.g7germany.de/ Webs/G7/EN/G7-Gipfel_en/Gipfeldokumente_en/summit-documents_node.html [accessed: 15.04.2015].

94. Owen, J.R./Kemp, D.: "Social Licence and Mining: A critical Perspective". In: *Resources Policy*, 38, 2013, pp. 29–35.

95. Parsons, R./Moffat, K.: "Constructing the Meaning of Social Licence". In: *Social Epistomol-ogy*, 28: 3–4, 2014, pp. 340–363.

96. van der Ploeg, F.: "Natural Resources: Curse or Blessing?". In: *Journal of Economic Litera-ture*, 49: 2, 2011, pp. 366–420.

97. The Economist: *The Dutch Disease*, 26.11.1977, p. 82–83.

98. Béland, L.P./Tiagi, R.: *Economic Freedom and the "Resource Curse" — An Emperical Analysis* (Studies in Mining Policy), Vancouver, Kanada: Fraser Institute 2009. URL: http://www.fraserinstitute.org/research-news/display.aspx?id=12972 [accessed: 15.01.2015].
99. Gwartney, J./Lawson, R./Hall, J.: *Economic Freedom of the World: 2014 Annual Report*, Vancouver, Kanada: Fraser Institute 2014.
100. Large, D.: "Neue Technologien in Exploration und Lagerstättenentdeckung". In: Kausch, P./Bertau, M./Gutzmer, J./Matschullat, J. (Eds): *Strategische Rohstoffe – Risikovorsorge*, Berlin, Heidelberg: Springer Spektrum Verlag 2014, pp. 149–158.
101. Bizzi, L.: "Minerals Exploration in Brazil: the Perspective of a major Mining Company". In: *BHP Billiton South American Minerals Exploration Group, ADIMB — Agéncia para o Desenvolvimento Technológico da Industria Mineral Brasileira Exploration manager's meeting*, April 2007. URL: http://www.adimb.com.br/eventos/gerentes/pdf/bizzi.pdf [accessed: 21.10.2014].
102. Schodde, R. C.: "Recent Trends in Copper Exploration — are we finding enough?". In: *International Geological Congress IGC Brisbane*, Australien, 05. – 10. August 2012), 2012. URL: http://www.minexconsulting.com/publications/IGC%20Presentation%20Aug%20201 2%20PUBLIC.pdf [accessed: 15.01.2015].
103. Buchholz, P./Liedtke, M./Gernuks, M.: "Evaluating Supply Risk Patterns and Demand Trends for Mineral Raw Materials: Assessment of the Zinc Market". In: Sinding-Larsen, R./Wellmer, F.-W. (Hrsg.): *Non-Renewable Resource Issues – Geoscientific and Societal Challenges,* Dordrecht, Heidelberg: Springer Verlag 2012, pp. 157–181.
104. Dorner, U./Buchholz, P./Liedtke, M./Schmidt, M.: *Rohstoffrisikobewertung – Kupfer, Kurzbericht* (DERA Rohstoffinformationen Nr. 16), Deutsche Rohstoffagentur in der Bundesanstalt für Geowissenschaften und Rohstoffe 2013.
105. Finlayson, E.: "Demand, supply and price of copper – an exploration perspective". In: *Rio Tinto, Global Mining Investment Conference*, London, 30. September 2009. URL: http://www.riotinto.com/documents/ReportsPublications/20093009__Finlayson__Glo bal__Mining_Investment_Conference.pdf [accessed: 21.10.2014].
106. Wagner, M. K. F.: "Ökonomische Bewertung von Explorationserfolgen über Erfahrungskurven". In: *Geologisches Jahrbuch Reihe H*: Heft SH 12, 1999.
107. Le Roux, C./Macnae, J.: "SQUID Sensors for EM Systems". In: Milkereit, B.: *Exploration in the new Millenium,* Toronto: Decennial Mineral Exploration Conferences, 2007, p. 417–423.
108. Schodde, R. C.: "The Key Drivers behind Resource Growth: an analysis of the Copper Industry over the last 100 Years". In: Paper Mineral Economics & Management Society (MEMS) session at the 2010 SME Annual Conference, Phoenix, Arizona, March 2010), 2010. URL: http://www.minexconsulting.com/publications/Growth%20Factors%20for%20 Copper%20SME-MEMS%20March%202010.pdf [accessed: 16.03.2014].
109. Scholz, R./Wellmer, F.-W.: "Approaching a dynamic View on the Availability of Mineral Resources: What we may learn from the Case of Phosphorus?". In: *Global Environmental Change* 23: 1, 2013, p. 11–27.
110. Schodde, R.C.: "Global Outlook and Development Trends for Copper". In: *Philippines Mining Conference*, Manila, 20.09.2012. URL: http://www.minexconsulting.com/publica tions/Copper%20Outlook%20-%20PMC%20presentation%20Sept%202012.pdf [accessed: 13.03.2014].
111. Weihed, P.: "Consequences of the RMI and the EU report on critical metals." In: *SDMI Proceedings*, Aachen 2011.
112. Schodde, R.C.: Personal telephone conversation, Melbourne/Kiel 10.9.2015.
113. Andruleit, H./Babies, H.G./Cramer, B./Krüger, M./Meßner, J./Rempel, H./Schlömer, S./Schmidt, S.: "Konventionell versus nicht-konventionell: Weltweite Ressourcen und Entwicklungen des,Hoffnungsträgers' Erdgas". In: *DGMK-Proceedings*, Celle, 12. – 13. April 2010.
114. Harris, M.: *Key note address*, SGA Meeting, 12. – 15.08.2013, Uppsala, Sweden 2013.
115. Hagelüken, C.: "Recycling 2.0 – Verbesserung des Recyclings wirtschaftsstrategischer Metalle durch systematische Optimierung u. industrielle Kooperation." In: *DECHEMA Info-Day "Wiedergewinnung strategischer Metalle"*, Frankfurt am Main, 13.05.2014.

116. Hagelüken, C.: "Technologiemetalle – Systemische Voraussetzungen entlang der Recy-
     clingkette". In: Kausch, P./Bertau, M./Gutzmer, J./Matschullat, J. (Eds.): *Strategische
     Rohstoffe – Risikovorsorge*, Berlin, Heidelberg: Springer Spektrum Verlag 2014, p. 161–172.
117. United Nations Environment Programme: *Recycling Rates of Metals – a Status Report*, Report
     of the Working Group on the Global Metal Flows to the International Resource Panel, 2011.
     URL: http://www.unep.org/publications/ [accessed 15.01.2015].
118. Kawohl, C.: *Recycling komplexer Materialzusammensetzungen*, acatech-Workshop Rohstoffe
     – neue Werkstoffe – neue Technologien, Berlin, 26.10.2011.
119. International Aluminium Institute: *Recycling Indicators*, 2014. URL: http://recycling.world-
     aluminium.org/en/review/recycling-indicators.html [accessed: 26.11.2014].
120. Bundesanstalt für Geowissenschaften und Rohstoffe: *Deutschland – Rohstoffsituation 2016,*
     Hannover 2017.
121. Steinbach, V./Wellmer, F.-W.: "Consumption and Use of Non-Renewable Mineral and Energy
     Raw Materials from an Economic Geology Point of View". In: *Sustainability*, 2, 2010,
     p. 1408–1430.
122. US Department of Energy: *Mining Industry Energy Bandwith Study*, Washington DC
     2007. URL: http://energy.gov/sites/prod/files/2013/11/f4/mining_bandwidth.pdf [accessed:
     09.10.2015].
123. Meskers, C. E. M.: Coated Magnesium—Designed for Sustainability? Dissertation TU Delft
     2008 (ISBN 978-9-064643-05-7).
124. Reuter, M.A./Verhoef, E.V.: "A dynamic Model for the Assessment of the Replacement of
     Lead in Solders". In: *Journal of Electronic Materials*, 33: 12, 2004, p. 1567–1580.
125. Verhoef, E.V./Dijkema, G.P.J./Reuter, M.A: "Process Knowledge, System Dynamics and
     Metal Ecology". In: *Journal of Industrial Ecology*, 8: 1–2, 2004, p. 23–43.
126. United Nations Environment Programme: *Metal Recycling – Opportunities, Limits, Infras-
     tructure*, Report of the Working Group on the Global Metal Flows to the International Resource
     Panel, 2013. URL: http://www.unep.org/publications/ [accessed 25.01.2016].
127. European Commission: *Critical raw materials for the EU* (Report of the Ad-hoc-Working
     Group on defining critical Raw Materials), Brussels 2010. URL: http://ec.europa.eu/enterpri
     se/policies/raw-materials/files/docs/report-b_en.pdf [accessed: 01.05.2014].
128. European Commission: *Critical raw materials for the EU,* Report of the Ad-hoc-Working
     Group on defining critical Raw Materials, Brussels 2014. URL: http://ec.europa.eu/enterp
     rise/policies/raw-materials/files/docs/crm-report-on-critical-raw-materials_en.pdf [accessed:
     26.06.2014].
129. Schebek, L./Becker, B.F.: "Substitution von Rohstoffen – Rahmenbedingungen und Umset-
     zung". In: Kausch, P./Bertau, M./Gutzmer, J./Matschullat, J. (Eds.): *Strategische Rohstoffe-
     Risikovorsorge*, Berlin, Heidelberg: Springer Spektrum Verlag 2014, pp. 3–12.
130. Graedel, T.E./Harper, E.M./Nasser, N. T./Reck, B.K.: "On the Materials Basis of modern
     Society". In: *Proceedings of the National Academy of Sciences of the USA* (Early Edition),
     2013. URL: http://www.pnas.org/cgi/doi/10.1073/pnas.1312752110 [accessed: 01.04.2014].
131. Tercero Espinoza, L./Hummeln, T./Brunot, A./Hovestad, A./Peña Garay, I./Velte, D./Smuk,
     L./Todorovic, J./van der Eijk, C./Joce, C.: *Critical Raw Materials Substitution Profiles*
     (CRM_InnoNet-The Innovation Network for Substitution of Critical Raw Materials), 2013.
     URL: http://cdn.awsripple.com/www.criticalrawmaterials.eu/uploads/Raw-materials-profile
     s-report.pdf [accessed: 25.11.2014].
132. Duclos, S./Otto, J./Konitzer, D.: "Design in an era of constrained resources". In: *Mechanical
     Engineering* 132:9, 2010, pp. 36–40.
133. US Congressional Budget Office: *Cobalt: Policy Options for a strategic Mineral*, Washington
     DC 1982. URL: https://www.cbo.gov/sites/default/files/cbofiles/ftpdocs/51xx/doc5126/doc2
     9-entire.pdf [accessed: 01.10.2015].
134. Wuppertal Institut für Klima, Umwelt, Energie GmbH: *KRESSE — Kritische mineralis-
     che Rohstoffe bei der Transformation des deutschen Energieversorgungssystems*, Abschluss-
     bericht an das Bundesministeriums für Wirtschaft und Energie, Wuppertal 2014. URL: http://
     wupperinst.org/de/projekte/details/wi/p/s/pd/38/ [accessed: 15.12.2014].

135. Bundesministerium für Wirtschaft und Energie: *Rohstoffstrategie der Bundesregierung – Sicherung einer nachhaltigen Rohstoffversorgung Deutschlands mit nicht-energetischen mineralischen Rohstoffen*, Berlin: BMWi 2010.
136. European Commission: *Tackling the challenges in commodity markets and on raw materials* (Communication of the Commission to the Council and the European Parliament), Brussels 2011.
137. Statistisches Bundesamt: *Umweltnutzung und Wirtschaft* (Bericht zu den umweltökonomischen Gesamtrechnungen 2014), Wiesbaden 2014. URL: https://www.destatis.de/DE/Publika tionen/Thematisch/UmweltoekonomischeGesamtrechnungen/Querschnitt/Umweltnutzungu ndWirtschaftBericht5850001137004.pdf [accessed: 18.02.2015].
138. Factor 10-Institute: *Manifesto*, 2010. URL: http://www.factor10-institute.org/files/F10_Man ifesto_e.pdf [accessed: 19.04.2014].
139. Bundesregierung 2002: *Perspektiven für Deutschland – Unsere Strategie für eine nachhaltige Entwicklung*. Berlin: Die Bundesregierung 2002. URL: http://www.bundesregierung.de/We bs/Breg/DE/Themen/Nachhaltigkeitsstrategie/1-die-nationale-nachhaltigkeitsstrategie/nach haltigkeitsstrategie/_node.html [accessed: 14.04.2015].

Chapter 4
Current Status of Natural Resources—An Overview

Two scenarios must be distinguished in an assessment of the supply of natural resources: the supply from domestic sources and the supply from foreign sources. Domestic natural resources, including both primary mining and secondary recovery of resources from end-of-life materials, are for industry the most reliable with respect to political interference. However, from a private business perspective, this distinction is irrelevant because there are contractual agreements between the exporter and importer. These supply contracts are usually also fulfilled if there are domestic shortages. Exceptions include political intervention such as export restrictions that are enacted by some countries, or sanctions that, for example, have been recently applied because of the Ukraine crisis.

Although the significance of mineral resource production in Germany has declined appreciably over the years, Germany remains an important mining country. Germany is the biggest producer of lignite and the fifth biggest producer of potash in the world, as well as being an important exporter of rock salt and potash (Fig. 4.1). Germany produces domestically nearly all the construction material resources (such as sand and gravel) as well as some of the industrial minerals (such as kaolin and gypsum) that are required for the home market. Of the non-renewable energy resources in 2013, 100% of the lignite, 13% of the hard coal, 12% of the natural gas and about 2% of the oil were produced domestically. The production of hard coal will terminate in 2018, and therefore most of the fossil energy resources must be imported. Fossil energy resources comprise more than two-thirds of the import balance for all resources (Fig. 4.2).

About 20% of the biomass used as an energy source in Germany in 2008 was imported [1], and about 80% was derived from domestic cultivation (see Sect. 4.3). The domestic biomass could only be used for generation of bioenergy because additional biomass was imported especially for animal feed. Germany is therefore very dependent on acquiring a proportion of the net primary production outside of its borders in order to meet the current total demand for biomass [2, 3]. Germany must also in future import an increasing proportion of the bioenergy fuels [3].

© Springer International Publishing AG, part of Springer Nature 2019   107
F.-W. Wellmer et al., *Raw Materials for Future Energy Supply*,
https://doi.org/10.1007/978-3-319-91229-5_4

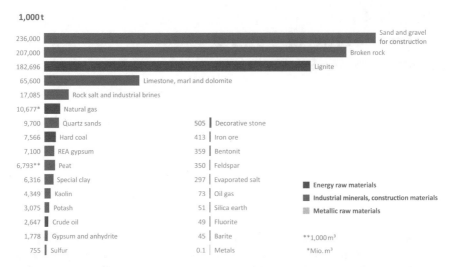

1,000 t

236,000	Sand and gravel for construction
207,000	Broken rock
182,696	Lignite
65,600	Limestone, marl and dolomite
17,085	Rock salt and industrial brines
10,677*	Natural gas
9,700	Quartz sands
7,566	Hard coal
7,100	REA gypsum
6,793**	Peat
6,316	Special clay
4,349	Kaolin
3,075	Potash
2,647	Crude oil
1,778	Gypsum and anhydrite
755	Sulfur

505	Decorative stone
413	Iron ore
359	Bentonit
350	Feldspar
297	Evaporated salt
73	Oil gas
51	Silica earth
49	Fluorite
45	Barite
0.1	Metals

■ Energy raw materials
■ Industrial minerals, construction materials
■ Metallic raw materials

**1,000 m³
*Mio. m³

Fig. 4.1   Natural Resource production in Germany in 2013 [14, p. 16]

Germany is practically one hundred percent reliant on imports of mined metallic mineral resources, in contrast to fossil fuels and biomass. The domestic secondary production fulfills only a part of the resource supply, and to a different degree for each raw material. It is therefore important for both Germany in general, but also for the energy systems of the future, that sufficient quantities of metallic resources are available on the international market. These are primarily supplied from primary mining operations. This includes chromite that is, for example, sourced from South Africa and Turkey, iron ores from Brazil, Canada and Sweden, germanium from China, Russia and the USA, as well as copper ores from Peru, Chile, Argentine and Brazil. The raw materials of strategic economic importance, small quantities of which have a significant leverage on the economy, are especially important. These include the steel alloy metals, refractory raw materials, raw materials for the electronics sector, and other high-technology raw materials such as rare-earth elements or platinum group elements. In 2013, their proportion of the total quantity of raw materials imported into Germany was 5.8% (Fig. 4.2).

As already discussed, the import of natural resources can be restricted by concentration effects at the country or company level (see Sect. 3.4.1) or by political intervention such as export restrictions (see Sect. 3.4.2), and these can lead to economic losses. From the perspective of German industrial concerns, these potential import obstacles represent risks for the free movement of raw materials and are therefore weaknesses in the industrial added-value chain. As a major industrial location, Germany could be affected by the following scenarios:

• International competitors have a possible advantage by better access to the sources of the resources, such that there could be a risk of German manufacturing compa-

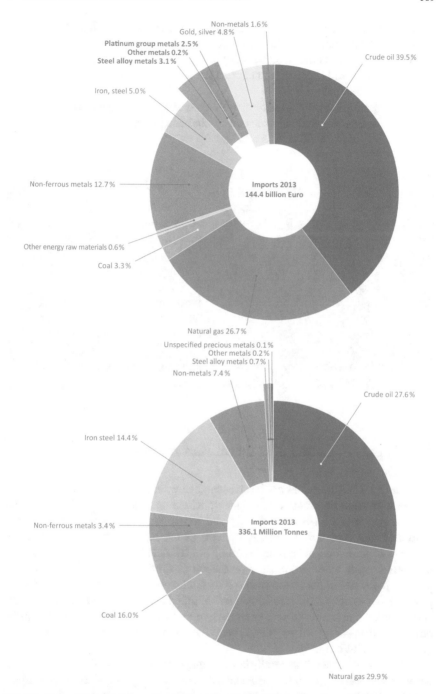

Fig. 4.2 Proportional value and quantity of the Natural Resources imported into Germany in 2013 (Updated and modified from BGR [14, p. 19]). The raw materials of strategic economic importance are emphasized

nies moving overseas. Export restrictions, as described above, have the effect of favoring the manufacturers in that country.

- Supply shortages could arise for certain raw material specifications, for which Germany is already very dependent on imports and for which there is only a low level of diversification.
- The competition on the international markets increases to the degree where the German companies find themselves in global competition, and in future the supplies of resources could be less reliable.

The current status of Natural Resources—an Overview
Mineral natural resources of metals, fossil energy resources and biomass are all important to varying degrees for the energy systems of the future. The metals are the most important group of resources for the energy system, based on the current technologies. Because the domestic secondary production of metals will not be able to meet the demand in the foreseeable future, Germany is dependent to a very high degree on the import of primary mined resources. Except for lignite, a large proportion of the fossil energy resources must also be imported. Although Germany is an important agricultural country, the country is not self-sufficient in the agricultural production of biomass for energy generation. Natural resources are sourced from countries in the whole world, and Germany is therefore dependent on the free movement of these resources.

4.1 Supply Situation of Mineral Natural Resources

The non-metal and the metallic natural resources are distinguished for this discussion.

The **non-metallic resources** include construction materials and the industrial minerals as well as fertilizer resources and water. The construction materials are obviously important for the energy systems of the future. Geologically they are available in unlimited quantities, but they are restricted by the competing claims on land-use such as environmental protection or drinking water. The availability of these resources in the future therefore is not an absolute problem, but a relative problem—or a problem of priorities. The unlimited availability is also the case if it is considered that the renewable energies require more resources at the investment phase, based on one unit of generated energy, as compared to the conventional energy generating facilities [4, 5].

The industrial minerals generally have a high potential for substitution. Apart from water, plants need the following elements, as fertilizer, to grow: nitrogen, potassium and phosphorus. These resources cannot be substituted. Nitrogen and potassium are present in unlimited quantities in, respectively, the air and seawater. This is not the case for phosphates that, with respect to their consumption, have very large reserves so that no shortages of the supply need to be expected up to 2050 and beyond [6]. Resources of freshwater are limited and should be reserved for drinking water and agricultural use.

Salt water is available in virtually unlimited quantities in the world's oceans and in arid and semi-arid regions, and can be desalinated to fresh water by an energy-dependent process.

The hydrogen in sea water contains 0.015% of deuterium (D_2), which can be enriched by various processes. Heavy water (D_2O) is required for nuclear fission reactors, and D_2 for nuclear fusion reactors, and both are available in almost unlimited quantities. With appropriate energy input, water can be broken down by electrolysis or photocatalysis into molecular hydrogen and molecular oxygen, both of which can also be used as sources for energy. Nonetheless, water is not an energy resource because the splitting of the water molecule requires more energy than is released by the burning of the molecular hydrogen. Molecular hydrogen, however, can be regarded as an energy resource but this is only present in trace amounts in the atmosphere.

Table 4.1 offers an overview of developments in the resources sector and provides the background as to why Germany is today one hundred percent reliant on imports for primary resources of **metallic raw materials**, which are the most important raw materials for the energy technologies of the future:

Because of their recent backward integration since 2003, some companies have successfully acquired interests in foreign mining in the metal sector. These companies are small as compared to the former Metallgesellschaft AG, which maintained a global trading network and could act as the contact organization for nearly every raw material. Metallgesellschaft AG was much more important not only in the mining and exploration sector, but also as a provider of engineering services.

The scenario in the sector of production from secondary resources is more encouraging. Today several German and European producers can meet a significant proportion of their raw material requirements from recycling, whereby the metal-bearing secondary materials are usually acquired from sources throughout the world. Examples of companies active in this business include Hydro for aluminum, Aurubis for copper and Umicore for precious and special metals.[1] European steel producers today process a large proportion of secondary materials. The development of more efficient recycling facilities could contribute to closing some raw material cycles, and therefore increase the reliability of supply.

---

[1]Umicore can meet a significant proportion of the demands for gold (for example for jewelry alloys and galvanic baths), for silver (for example for contact materials), and for platinum metals (for catalysts) from the recycling of electronic fractions, catalysts and industrial waste. Additionally, the company uses in-house recycled germanium for opto-electronic products or secondary cobalt for carbide and battery materials.

Table 4.1 Economic and political events in the German mining sector from 1970 to 2013

Period	Events
Prior to 1992	The Federal Government's program to support exploration ("Measures to improve the supply of mineral natural resources to the Federal Republic") from 1970 to 1990 succeeded in supporting German companies to acquire important shares in foreign mines and therefore develop a robust backward integration (See glossary) of their business [64]. The last metal mines in Germany closed in 1992. Until that time, the Federal Republic of Germany maintained significant self-sufficiency in some primary raw materials, for example zinc and lead with germanium as a by-product. The deposits were either exhausted or the grades were too low to remain competitive. The metal prices in 1992 (The USD is used throughout the natural resource sector. The increasing value of the German Mark resulted in lower income for the German mines that were forced into a cost squeeze) were too low for a profitable operation. The copper and tin mines in the GDR were forced to close after the reunification because of their unprofitability. Therefore, from this moment in time Germany could only rely on the domestic production of metals from secondary sources
1992–2003	Following the general trend of the 1990s in Germany to rely only on the commodity markets, and after the total restructuring of Metallgesellschaft AG (from 1993) which was one of the most important German companies in the metal mining and mining sector as well as being a national icon for the supply of metal resources [65, 66], almost all the German shares in foreign mining operations were sold, often to direct competitors. A magnesite deposit in Canada is the only example of the exploration successes resulting from the exploration support program from 1970 to 1990 that has remained in German ownership. The closure of the metal mines in Germany, the relinquishment of German participation in foreign mines as well as termination of practically all exploration activities has resulted in a very significant loss of German expertise in the exploration and mining sector. This is also reflected in the university education. The trend to convert numerous departments of ore deposit research into the more modern "environmental geosciences" is an international phenomenon, even in traditional mining countries such as Canada and Australia
From 2003	Germany began to rethink its natural resource strategy only after the strong growth in global resource consumption as well as the continuously increasing competition from China on the international commodity markets
From 2009	The resource research programs of the BMBF started within the Program for Research on Sustainable Development (FONA as well as the high-technology strategy of the Federal Government [67] r^2 Innovative Technologies for Resource Efficiency—resource intensive production processes [68] r^3 Innovative Technologies for Resource Efficiency—strategic metals and minerals [69] r^4 Raw Materials of Strategic Economic Importance [70]

(continued)

Table 4.1 (continued)

Period	Events
From 2010	The Federal Government formulated its resource strategy [71]. The most important actions are 2010: Formation of the German Mineral Resources Agency (DERA) at the Federal Institute for Geosciences and Natural Resources (BGR) 2011: Formation of the Helmholtz Institute Freiberg for Resource Technology 2011: Adoption of the National Masterplan of Maritime Technologies NMMT by the Federal Ministry for Economic Affairs and Technology (BMWi): Germany, High-technology-base for marine technologies for the sustainable utilization of the oceans 2012: Formation of the "Alliance for Securing Natural Resources" by German industry, inspired by the Federation of German Industry (BDI). Due to a lack of interest by industry, this alliance is de facto no longer functional since the end of 2015 2012: Adoption of the German Resource Efficiency Program ProgRESS by the Ministry for Environment, Nature Conservation, Building and Nuclear Safety (BMUB) [72] 2013: Restart of the exploration support program by the Federal Ministry for Economic Affairs and Energy [73]. This program was discontinued on 30 June 2015 due to lack of demand from the German industry

4.1.1 Founding of a German Natural Resource Company?

The possibility of establishing another company such as the Metallgesellschaft AG should be discussed with respect to the supply of natural resources to German industry. This is impossible before 2020, and is questionable before 2050. There are no longer any domestic metal ore mines that could act as a basis for expertise. The trading network of Metallgesellschaft was developed over decades of cultivating contacts, some of which had survived the two World Wars, but now no longer exist. The first foreign successes of the Metallgesellschaft AG were developed through this network. Although the RA Resource Alliance GmbH of the German industry was active from 2012 to 2015, it was not successful due to the lack of interest from the industry. The rapid development of a new internationally meaningful German mining company, which would be active anticyclical, probably requires the initial support from the state, such as was practiced in the case of the oil sector with the Deminex[2] organization, and today initiated in France with the Compagnie Nationale des Mines de France.[3] Other relevant examples include the successful start-ups of

---

[2]Deminex was a company founded by the German oil companies, for which the Federal Government provided 2.375 billion German Marks of federal funds between 1969 and 1989 as loans and subsidies for the securing and improving the supply of oil for the Federal Republic of Germany [7].

[3]For example, Le Parisien 21.02.2014 [74]; FAZ 21 February 2014 [75]; Die Welt 21 February 2014: In February 2014 Minister for Industry Arnaud Montebourg announced the formation, in cooperation with private industry, for a state-directed mining company Compagnie Nationale des Mines de France. Latter initiative was, however, cancelled by the French government in 2016—without the

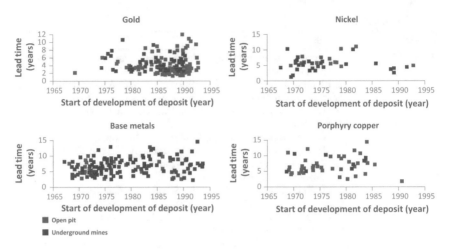

Fig. 4.3 **Lead times (Lead-Time-III) for the development of a project from the final explo-**
ration activities to the commencement of production for various raw materials [61]. "Porphyry"
relates to large tonnage, low grade copper (±molybdenum and gold) deposits

Kupferexplorationsgesellschaft mbH[4] (copper exploration company) or the German
Urangesellschaften[5] (uranium companies), by which the raw material customers,
encouraged by the Federal Government, cooperated with strong mining companies
to secure raw materials supply. These companies contributed to supplying German
industry in the copper and nuclear energy sectors; Their successes would not have
been possible without strong mining companies with the appropriate expertise (from
mining to marketing).

BOX XIII Lead Times during Supply Shortages

If there are shortages of natural resources then the prices increase. During
these times of higher prices, it is possible to develop deposits that were already
known but were not economic to mine during low price periods. These deposits
are referred to by experts as "shelved deposits". The International Strategic
Minerals Inventory (ISMI) working group from the geological surveys of

---

company having been founded. https://www.challenges.fr/energie-et-environnement/macron-enter
re-la-compagnie-des-mines-de-france-chere-a-montebourg_40701.

[4]Under the direction of Metallgesellschaft AG in 1978, the Degussa AG, Siemens AG and Kabel-
und Metallwerke Gutehoffnungshütte AG (each with a 25% share) combined into the Kupferexplorationsgesellschaft mbH.

[5]See Anger [8]: there were three German Urangesellschaften: Urangesellschaft mbH (Metallgesellschaft AG, VEBA and STEAG), Uranerzbergbau-GmbH (at that time RWE-daughter Rhein-
braun, and C. Deilmann AG) as well as Saarberg Interplan Uran GmbH (Saarbergwerke AG,
Badenwerk AG and Energieversorgung Schwaben AG).

Australia, Germany, Canada, South Africa, the USA and Great Britain have compiled the so-called Lead-Time-III-Study [9]. "Lead-Time-III" refers to the time that is required to bring a deposit, either a newly discovered deposit or a shelved deposit, into production with no interruptions.

Lead-Time-III times are dependent on the raw material being mined. During the 1990's the shortest lead times were for gold with an average of three years, and for porphyry copper deposits the lead times were an average of seven years, whereby there is a considerable range (Fig. 4.3). The longest lead times were almost thirty years. The comparison of the large open pit projects (porphyry copper deposits), which are especially cost intensive, with other raw materials clearly shows that the length of the lead times increases with increasing capital cost requirements. The increasing size and complexity of the mine facilities also results in longer lead times, so that in 2012 lead times of five to twelve years are reportedly required for rare-earth elements in comparable situations [10]. These times are almost exactly the same for the metals shown in Fig. 4.3, but unfortunately the study has not been updated.

Although neither Fig. 4.3 nor the results of other studies [11, 12] provide any indication of a significant trend towards longer lead times, increased lead times may be expected in the future due to the stronger regulations for the environmental and social issues. It can be concluded that the period of supply shortages, which is the time required to bring new or shelved deposits into production, is currently limited to about ten years. The political dependability of the producer country is another issue. Current lead times of eleven (low) to nineteen (high) years, depending on the country risk, for copper deposits.[6]

Bearing in mind that lead times are required before production can commence from a newly discovered mineral deposit, the duration of state intervention or support is critical to the future supply of natural resources. During periods of normal supply and demand relationships the lead time is about ten years. The lead time is longer in risk-prone countries as compared to more stable countries (BOX XIII). From the perspective of the market economy, the start-up phase should be as short as possible. In order that industrial mining expertise is once more readily available in the German metal sector, the possibility of a joint venture with an international, European-controlled, metal mining company is worth considering as a concept that could possibly be realized at the European level. It is obviously necessary to ensure that the mother company does not compete with the joint venture. Eramet, which is a French mining and processing company operating nickel mines on New Caledonia and manganese mines in Gabon that also supports its own research department, is an example of a suitable candidate as a partner company.

---

[6]Schodde [12]; lead times are for copper deposits.

4.1.2 Avoidance Strategies for a Reliable Supply

The backward integration with mine production is the most reliable, but only one of several strategies for companies to ensure a reliable supply of raw materials. Other strategies could be:

- Concluding long-term supply contracts with price escalation clauses and hedging transactions[7]
- Form purchasing groups
- Diversify the supply sources (various suppliers and countries)
- Sufficient stockpiling
- Increase the material efficiency and improve the in-house recycling
- Increase the flexibility within the supply chain by substitution and escape clauses, and by discussing in advance with the customer the possibilities to change to other raw materials.

4.1.3 Secondary Resources Increase the Reliability of Supply

Secondary resources are increasingly important with respect to the reliability of supply. Basically, the potential for the supply from secondary resources rises with increasing resource inventory in infrastructure and products. As discussed in Sect. 3.4.4, there is a continuing shift from geological to anthropogenic deposits. The faster that the product cycles change, and the shorter the lifetime use of resource relevant (mass)-products, the greater the contribution from secondary resources to the supply can become. This is already the case for many metals. However, the domestic secondary production cannot cover the demand, especially for an export-oriented national economy.

The use of secondary resources, conditional upon the overall situation such as investment security, has the following advantages in Germany and Europe over the supply from primary resources:

- The lead times for activating production from the currently unused potential of secondary deposits may be expected to be relatively short. The investment requirements are also relatively low.
- The public acceptance for recycling projects is better than for mining.

---

[7]The term "hedging" describes a financial transaction to secure a contract against risks such as exchange rate fluctuations or changes in the commodity prices. The company that wishes to hedge a transaction makes an additional offsetting deal that is back-coupled with the original transaction. This is normally in the form of a forward transaction.

- Germany and Europe have the most advanced development as well as extensive utilization of modern recycling technologies in the world, for both the mechanical processing as well as the metallurgical recovery of metals. Although exploration and mining expertise in Germany has been lost because of the decline in the domestic mine production and the withdrawal of major mining companies, the reverse trend can be recognized in the recycling sector: for the process development, engineering, plant construction as well as the operation of facilities for the recovery of metals from secondary resources.
- The leading German and European companies that are active in the recycling business are, to a certain extent, the successors of the major mining companies in their role of supplying natural resources. This is not limited to the domestic secondary resources, because already today the leading European metal smelters are importing secondary resource materials from sources around the world.

Because of the very high dependence on imports of metallic resources, accessible and open global markets are essential for the future availability of natural resources in Germany, so that a customer and consumer country such as Germany can always apply its interests in normal and undistorted competitive markets. The supply of natural resources can then be regarded as secure in the long-term. An exception could include extreme seller's market conditions, such as the oil crises of 1973 and 1979. Generally, however, the customer has the choice of where the resources are purchased insofar as it is not a monopoly. It is obviously easier to secure the supply of natural resources if there are numerous suppliers, and not as is the current overseas trade in iron ore that is an oligopoly dominated by a few suppliers. The degree of concentration is therefore an important parameter for defining critical resources (see BOX IV). Despite this, the persistent principle that markets always swing between the sellers and buyers continues to prevail, a seller must find a customer for its products, as natural resources in a mineral deposit are worth nothing if they cannot be sold. The Minister for Mining and Energy of Newfoundland, a Canadian province, formulated this very clearly in a speech in October 2001 with respect to the conditions for the development of one of the largest nickel deposits in the world at Voisey's Bay, when he said: "We must balance the need of the province to realize the maximum benefits for our shareholders, the people of the province, with the need of Inco to earn a fair return for its shareholders." Later in his speech he continues: "it goes without saying that our natural resources benefit no one if they remain undeveloped" [13]. In this context, political sanctions, such as those until recently against oil and natural gas exports from Iran, can be very important.

The Supply Status of Mineral Natural Resources

The domestic production capacity for primary metallic natural resources has been completely shut down in Germany because of the economic conditions or because of the depletion of the mineral deposits over time. German companies have also relinquished nearly all their shares in foreign companies or deposits because of the sufficient supply of natural resources. The relevant industrial expertise has therefore also been lost. However, the increases in demand and prices since the beginning of this century have resulted in several recent political (e.g., on research programs) and industrial initiatives to strengthen the domestic metal sector again and to improve the supply situation. An internationally significant backward integration that is necessary for greater independence from the resource markets has partly materialized for secondary resources, but not for primary natural resources. German and European producers source their requirements for secondary resources from the whole world, which meets most of their demand. Initial state support is probably required for a renewed backward integration in the primary natural resource sector. In order that industrial mining expertise is once more readily available in Germany, the possibility of a joint venture with an international, European-controlled, metal mining company is worth considering as a concept that could possibly be realized at the European level.

Non-metallic natural resources such as construction material, industrial minerals or water are, with respect to their supply, mostly not critical for the energy systems of the future. However, the energetic and environmental issues must be factored into their utilization.

4.2 Supply Situation of Fossil Energy Fuels

The German Federal Institute for Geosciences and Natural Resources (BGR) regularly issues studies on the reserves, resources and availability of fossil energy fuels [14]. This book "Raw Materials for Future Energy Supply" uses the 2014 study as a basis. Most of the data in these studies on the global situation are derived from reports of the International Energy Agency and other organizations [15, 16, 17]. The most recent studies on oil and natural gas, in contrast to earlier years, now also take the so-called unconventional fuel deposits into consideration. These refer to the oil and natural gas that occurs in very dense rocks, usually sedimentary rocks such as claystones with very low permeability, and usually can only be extracted with additional technical measures (hydraulic stimulation or fracking), as well as natural gas in coal seams (coalbed methane) and in methane clathrates, in which methane is trapped in the crystal structure of water and occurs in permafrost soils and in the deep ocean.

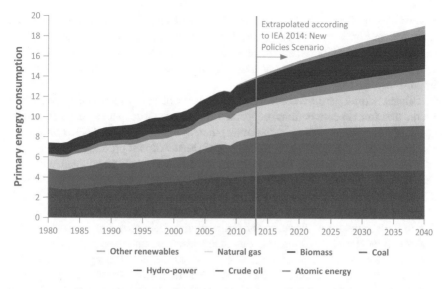

Fig. 4.4 Global energy consumption from 1980 to 2013, and extrapolated to 2040, in giga-tonnes oil-equivalent and subdivided according to the energy fuels (From BGR [14, p. 14]; extrapolation after IEA [15]). The International Energy Agency forecasts an additional increase to 2040, mainly due to increasing consumption in China and various other emerging countries (New Policies Scenario of the IEA)

According to these studies more than eighty percent of the primary energy consumption is from the fossil fuels oil, natural gas and coal (Fig. 4.4). Biomass, hydropower, and nuclear energy are also significant with a combined contribution of twenty percent. The International Energy Agency forecasts a continuing increase in the energy consumption, which is not due to growth in western Europe but primarily to growth in China as well as several emerging countries.

The configuration of the fuel sources required to meet the energy consumption in Germany is very similar to that internationally. About 80% of the energy is sourced from fossil fuels, about 10% from atomic energy and 10% from sources of renewable energy [18, p. 4]. However, the consumption in Germany has decreased slightly since 1990, and a further reduction is included in the objectives of the energy policies of the Federal Government.

The expected increase in global energy consumption also suggests a shortage of fossil energy fuels in the future, although the situation is different according to the type of energy fuel. The increase in the consumption of coal or atomic energy seems not to be critical with respect to the availability of these resources, because there are existing large resources and reserves of hard coal,[8] lignite and uranium. Uranium is not an issue for Germany because of the withdrawal from atomic energy planned by 2022. Practically all the lignite is derived from domestic production, but

---

[8]This does to apply to special coals, such as that required for the production of coking coal.

the proportion of domestic hard coal has declined drastically over the past ten years; it contributed only 13% in 2013, and the trend is continuing downwards. This is related to the termination by 2018 of subsidization of the German hard coal, which will also result in even less use of domestic hard coal and a greater dependency on imports after 2018.

Oil and natural gas are scarce natural resources, as compared to coal, even if the production from unconventional deposits (for example those that can be exploited by fracking processes or from gas clathrates) increases significantly. There continue to be controversial discussions on the timing of the maximum production from either of these two fossil energy fuels. It is assumed that natural gas and oil will continue to make a major contribution to the global energy consumption over the next 30–40 years, and that natural gas and oil extracted from unconventional deposits will increase. Despite the recent fall in the prices for oil and natural gas, which is partly due to the successful exploration for shale-oil and shale-gas, it is forecast that the prices will increase in the long term due to increasing demand and/or higher production cost, because of technical challenges with respect to e.g. deep sea production or unconventional shale oil and shale gas production. However, it should be noted that the price trends of the past were not only related to production costs but, to a considerable extent, also to political events, for example in the Middle East.

Technological developments in the exploration and production sectors of the industry have had a positive effect on price trends. These advances include modern techniques such as 3D-seismics, numerical modelling, horizontal drilling and hydraulic stimulation in the bore holes (fracking). The current significant reduction in the oil and natural gas prices is mainly due to the increased production from the unconventional deposits of shale oil and shale gas in the USA that were first exploited by fracking. The geological situation of these deposits are being investigated and evaluated, for example by the US Environmental Protection Agency (EPA), for their possible impacts on the environment [19]. These issues have also been evaluated in Germany in several different studies: including by the BGR [20, 21], by a group of independent experts commissioned by ExxonMobil Production Deutschland [22], on behalf of the Federal Agency for the Environment [23], on behalf of North Rhine-Westphalia [24] as well as by acatech, which is the German National Academy of Science and Engineering [25]. These technological advances could counteract an increase in the prices for oil and natural gas in the future. These include the coal liquefaction or gas-to-liquids (GtL) technologies from which fuels can be produced from natural gas.

Germany, with respect to natural gas and oil, is very dependent on a few source countries. Natural gas from domestic deposits met about 12% [14] of the German total production in 2013. Some of the natural gas is today sourced from the Netherlands, but as a result of declining reserves the Netherlands will no longer be an important supplier. Germany will become increasingly dependent on Russia for its supplies of natural gas unless unconventional deposits (for example by application of fracking technologies) are developed or natural gas is sourced from other continents. One solution would be the increased utilization of liquefied natural gas (LNG) that can be transported by tankers from terminals in other countries. There are fundamental

Table 4.2 Global production, estimated value of the global production in 2015 (Calculated from BP [17]) and German imports of fossil energy fuels in 2013

	Global production	Value (estimated) in billion €	German imports in billion €
Oil	32×10^9 bbl	2429[a]	55.98
Natural Gas	3.5×10^{12} m^3	346[b]	37.84
Coal	8.1×10^9 t	405[c]	4.71

From BGR [26]

[a]at 75 €/bbl

[b]at 0.1 €/m^3, the natural gas price can vary according to regional markets, which is different to that for oil

[c]at 50 €/t, coal prices are very variable, depending on type of coal (hard coal is much more expensive than lignite) and regional markets

strategic and economic political issues arising from the situation pertaining to the future supply of these natural resources.

The proportion of the domestic production of oil was only about two percent in 2013 [26], and thus much less than that for natural gas. Nonetheless there have been some surprising new discoveries of oil in Germany, such as that Römerfeld field near Speyer. Oil is the most expensive of the energy fuel resources. In 2013 Germany, which is the sixth largest importer of oil in the world, sourced its oil mainly from Russia, Norway and the United Kingdom as well as a significant proportion from politically less stable countries in the Middle East and North Africa,[9] but the production from the North Sea will decline in the foreseeable future. From the perspective of the national economy and strategic issues, it would be reasonable to limit the use of oil to those sectors where it cannot be easily substituted, for example as a chemical raw material or for mobility.

Sudden price fluctuations are typical for the price trends, for example because of shortages caused by OPEC production actions, economic crises such as in 2008, or political and military conflicts (Kuwait crisis and Iraq war). The price increased 10 times between 2000 and 2008 from 10 USD/barrel to over 100/USD/barrel. Since then the oil price has fluctuated between 40 and 70 USD/barrel. In 2013, the value of imported fossil energy fuels was 55.98 billion Euros for oil, 37.84 billion Euros for natural gas, and 3.3 billion Euros for hard coal. This comprises more than two-thirds of the value of all German imports of natural resources [14, p. 19], and emphasizes the importance to the national economy (compare Fig. 4.2, Table 4.2).

Methane clathrates from the deep oceans are another, but today uneconomic, source for hydrocarbons. The gas is contained in the crystal lattice of ice-like structures. The gas occurrences are considered to be very large, although these estimates are associated with great uncertainties. The information on the available quantities of

---

[9]BGR [14]: The proportion of crude oil imports to Germany from Russia in 2013 was close to 35%, followed by Norway with just over 12% and the United Kingdom with just over 10%. Nigeria, Kazakhstan, Libya, Azerbaijan, Algeria, Saudi-Arabia and Egypt all contribute between one and eight percent each. These countries together supply over 91% of the German imports.

clathrate gases is imprecise because, until now, only occurrences have been quantified and not economically recoverable reserves. Production from deep marine clathrate occurrences would seem to be improbable in the near future.

However, gas clathrates also occur onshore in permafrost regions, and methane gas is already being exploited from these occurrences such as the Mesoyake field in Siberia. In the short-term however, the extraction of natural gas from coalbeds is much more relevant. There are production possibilities for this gas in Germany, especially in North Rhine-Westphalia, that must be further evaluated with respect to the quantities of gas and feasibilities for extraction.

Even if electricity generation from wind power and photovoltaic continues to be quickly expanded, coal and natural gas powered generating facilities must continue to be kept available so long as no long-term storage of electricity is possible. The ensured power supply must be covered by these traditional technologies during periods when there is no sunshine or the wind is calm. The question of whether coal and natural gas are used depends on numerous issues. On the one hand coal is available in the long-term and is relatively inexpensive, and has the additional advantage for the national economy in that it can be produced from domestic deposits, and in the case of lignite the costs are break-even. Coal powered generating facilities have the disadvantage that they cannot be quickly fired up or shut down, as would be necessary to cover the fluctuations in the amounts of solar or wind electricity that are produced. The start-up phase takes several hours. Furthermore, coal power stations should be operated at a minimum of the forty percent of the rated capacity, as only then can they efficiently provide the energy to cover the fluctuations in the electricity production from solar and wind power facilities.[10] Finally, as compared to all the other energy fuels coal power produces the most CO_2 emissions.

Natural gas power stations can, in contrast, be quickly fired up and can flexibly supply energy. They are therefore well suited to offset the fluctuations in the input of electricity generated by wind power and photovoltaic facilities. Furthermore, the burning of natural gas is much cleaner as compared to that of coal. In the long-term, however, natural gas is less abundant and more expensive as compared to coal.

Despite the accelerated development of the electrical power facilities based on renewable resources and the higher impact of emissions derived from the use of fossil energy resources, the fossil fuels will continue to be required in the long-term. Effective systems to store renewable energies must be developed if the requirements for coal and natural gas in electricity generation are to be reduced. Underground storage facilities as well as surface pump systems have by far the greatest storage capacities. National energy reserves, either as domestic deposits or artificial storage of oil and natural gas, are very important for the security of supply. Not only can this offset any shortages in the supply, but can also act as a buffer against the associated periods of high prices. The resource markets are characterized by significant price fluctuations caused by relatively small shortages in supply. It is therefore important

---

[10]This is the case for today's power stations that are designed for permanent operation. Experts consider that coal power stations can be designed to be much more flexible in the future, for example by reducing the minimum load [27].

to retain the control over the amounts required to offset shortages during peaks in demand. Germany can control about one third of the natural gas supply in that 12% of the demand is covered by domestic production and 24% derived from the national natural gas storage facilities. The gas that is stored for long periods in underground storage facilities is used, for example, to meet peak demand in cold winters.

As the proportion of fluctuating energy supplied from wind or photovoltaic increases, so could importance of gas storage also increase. Surplus wind and solar energy can be converted by so-called power-to-gas processes into hydrogen or artificially produced methane, and then stored in large gas storage reservoirs. This sort of long-term storage is, apart from bioenergy, the only possibility to manage periods lasting several weeks with calm wind and low sunshine levels, without using the traditional fossil fuels [28]. Public acceptance is also important for this issue. Underground storage is generally more acceptable (locally) than the pumped-storage power plants at surface that also have fewer possible locations, and this is a major advantage. However further research on storage technologies is required to avoid accidents such as that which recently occurred near Gronau[11] in Germany.

> **Supply Situation for Fossil Energy Fuels**
>
> Fossil energy resources are currently an essential basis of our energy supply. This demand will decline with the increasing deployment of renewable energy technologies. A decoupling of the energy system from these energy resources is only conceivable once sufficient possibilities for the storage of energy are available to meet the demands of the fluctuating supply of solar and wind energy. Until then, the fossil energy fuels will remain essential for the energy supply.
>
> The import dependency of Germany on fossil energy fuels will continue to increase without either exploiting unconventional oil and natural gas occurrences or a technological and unsubsidized revival of the hard coal mining. This involves the danger of being even more reliant on individual, politically unstable, countries such as Russia.
>
> From the perspective of the national economy and strategic issues, it would be reasonable to limit the use of oil to those sectors where it cannot be easily substituted, for example as a chemical raw material or for mobility.

---

[11] In February 2014 leakage occurred from a 1000-m-deep storage facility system in a salt cavern. After this leak was assessed and limited operation of the storage reservoir was restarted, stored oil was discharged on the surface that resulted in significant environmental impacts.

4.3 Supply Situation for Biomass

Plant biomass is the most important resource for bioenergy. There are various matters that influence the amount of biomass that is available: the light irradiation, the useful areas under cultivation, the soil types, the availability of water as well as the fertilizing with nitrogen, phosphate and potassium. To achieve the best possible yields of biomass, the arable land, grasslands and pastures must be worked, fertilized and sometimes irrigated. Only the forest areas are not usually fertilized with additional minerals. Globally, the quality of the soil and the availability of water for agriculture are limiting factors. An imminent phosphorus peak is repeatedly discussed, but an actual limitation is currently not in sight.

Much of the information and formulations used in this chapter are derived from the statement on *Bioenergy—Chances and Limits*[12] from the National Academy of Sciences Leopoldina that was published in 2013. The information on the greenhouse gas balance of renewable biomass is also derived from this publication. The following general conclusions from this study include: the greenhouse gas balance of sustainably managed forests is mostly neutral. By way of contrast, the greenhouse balances of intensively cultivated arable land, grasslands and pastures are positive on the basis of the net production of the greenhouse gases carbon dioxide (CO_2), methane (CH_4) and nitrous oxide (N_2O) (Table 4.3).[13]

In 2011, almost eight percent of greenhouse gas emissions in Germany were derived from agriculture [30: Federal Environment Agency]. Globally this figure is about twenty percent. The use of agricultural biomass as a source for energy is therefore not sustainable with respect to the climate, even if less greenhouse gases

Table 4.3 Greenhouse gas emissions related to the growth of biomass expressed as a percentage of the carbon dioxide assimilated (CO_2) in the harvested biomass [2, pp. 26–28]

Farmed areas	CO_2 from soil-carbon (%)	CO_2 from fossil fuels (%)	Greenhouse gases from harvest residues (%)	N_2O and CH_4 (Emissions due to the use of fertilizers) (as CO2 equivalent) (%)	$\sum$ (%)
Arable land	4	11	14	12	41
Meadows and pasture	−26	7	18	20	19
Woodlands	−32	3	21	1	−7

The data relates to the EU-25 countries. The error range is at least ±10%. Negative figures relate to the absorption of CO_2/greenhouse gases out of the atmosphere and positive figures indicate an emission of CO_2/greenhouse gases into the atmosphere

---

[12]Leopoldina [2]: based on the English version of the statement "Bioenergy—Chances and Limits" that was published in 2012 [29].

[13]Leopoldina [2, p. 27]: "N_2O and CH_4 have a much stronger greenhouse effect as compared to CO_2. The potential of CH_4 is about 25-times and that of N_2O is about 300-times higher than that of CO_2, relative to a baseline one hundred years ago."

are emitted as compared to the use of fossil energy fuels per energy unit. Intensive agriculture leads to a decline in biodiversity, water consumption and contamination of waters with the excessive use of nutrients [31]. This also contributes to the observation that the utilization of such biomass for energy generation is not sustainable.

The annual rate at which mankind has been exploiting many of the natural resources of the Earth is now gradually declining after many years of continual increase. For example, this includes the decline in the conversion of meadows and forest to arable land, in the increase of irrigated agricultural fields, the water consumption, and the increase in yield per area unit, while the rate of consumption of gas, oil and coal, but also of phosphate, is continually increasing. It is interesting to note that 16 out of 20 independent renewable resources reached a peak global consumption-rate at about 2006, including for example the consumption of rice, sugar cane, soya, grain and wood. The rate of increase of the global population had already peaked in 1989 [32].

4.3.1 Proportion of Bioenergy in Primary Energy Consumption

The primary energy demand from industry, transport and households of the about seven billion people on earth is about 560 EJ[14]/year. This demand for primary energy is met by fossil fuels, atomic energy, and from renewable sources such as renewable biomass. The annual contribution of bioenergy to the global primary energy consumption is about 55 EJ, or nearly 10%[15] (Fig. 4.5). Bioenergy, especially wood, is the main source of energy in most of the developing countries.

The primary energy consumption in the Federal Republic of Germany was 13.75 EJ in 2013, and 13.08 EJ in 2014. The difference is due to the mild weather during 2014. In 2014, bioenergy contributed a little more than 0.98 EJ (7.5%) to the total consumption. Over fifty percent of the biomass used in energy production, primarily wood, is for heating. About 25% was used for electricity production and about 16% was used as fuel. Biodiesel is the leading type of bio-fuels [18, p. 39ff]. In fact, practically all of the biomass that is harvested in Germany is required for nutrients, animal feed as well as construction materials. Only the residues from biomass remain for conversion into bioenergy. The contribution of 7.5% to the total bioenergy is only possible by significant imports of biomass. Imports of biomass often result in direct or indirect shortages of biomass elsewhere for nutrition, results in deforestation of rain forests, or the decline of biodiversity.

---

[14]One Exajoule is one trillion or 10^{18} J.

[15]This is equivalent to 1.5 billion tonnes of carbon per year.

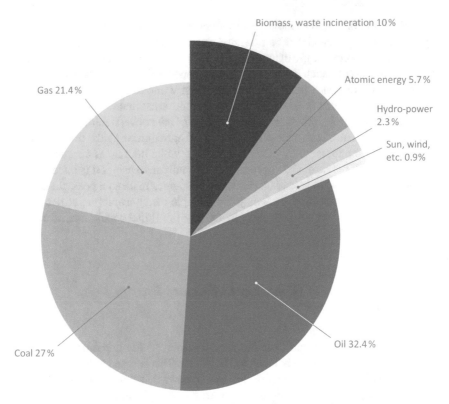

Fig. 4.5   Global primary energyconsumptionin 2012 (560 EJ), subdivided according to energy sources (Data from IEA [62, p. 6])

The importance of biomass for the energy supply in Germany is emphasized by the example of electricity. At the end of 2013, biogas was produced in about 8000 biogas facilities [33] and was continuously converted into electricity with an efficiency of up to 35%. The installed electrical capacity of all these facilities together is about 3.54 GW, and the net production is about 2.5 Gigawatt. The total electricity generated in 2013 is reported to be 27.5 TWh, which is about four percent of the total electricity generated in Germany in 2013 (633 TWh). Taking only the German generation of electricity from renewable energy sources (152 TWh) into account, then biogas contributes approximately 18% to the Eco-Electricity [18, pp. 30, 39]. These figures emphasize that the demand-oriented production of electricity from biogas is quantitatively important during those periods, in which electricity from wind power or photovoltaic is not available.

The global annual consumption of primary energy per head is currently about 70 GJ per person and year. This global annual average is much less than the average in Europe (150 GJ per person and year) or in the USA, Canada, Norway and Australia (>300 GJ per person and year). Since the standard of living of the people in other countries will hopefully improve, it must be expected that the global average will increase during the next 40 years to at least 100 GJ per person and year. In addition, the total global population is expected to increase from 7 billion to 9 billion during the same period. It is therefore very probable that the annual global energy consumption will almost double from the current 560–900 EJ by 2050. This increase could be less if the primary energy consumption in the industrialized countries can be reduced. Germany has set itself an objective by 2050 to half the current level of almost 14–7 EJ/year. However, during the last few years there are scarcely any indications for a meaningful reduction in the energy consumption.

The global consumption of biomass will increase significantly together with the consumption of primary energy because additional amounts will be required for food for human consumption as well as the manufacturing of products such as paper, cotton and construction materials. Considering that in the future agriculture should have less negative effects on the climate and environment, it is clear that the available amount of biomass will not significantly increase. The proportion of bioenergy in the global consumption of primary energy will probably decrease from the current level of 10%.

Finally the forecasts on the global potential of bioenergy are uncertain because they are dependent on several parameters:

- The expected increase of the harvest yields
- The sustainable amount of available water
- The greenhouse gas emissions that are associated with intensive agriculture practices
- The effects of direct and indirect changes in the use of the land on the greenhouse gas emissions and the biodiversity
- The areas used for agriculture that are required for the supply of nutrition to humankind.

Obviously some of these parameters will be affected by the level of population, the degree of affluence, the losses in the food manufacturing chain, and the eating habits of people [34]: For example non-vegetarians require a much larger agrarian area for their food as compared to vegetarians [2, p. 33f]. The estimates in the current "forecast" by the Club of Rome [35] differ from those by the International Renewable Energy Agency (IRENA) [36]. Both estimates assume a significant future increase in the biomass available for nutrition and bioenergy. The "forecast" from the Club of Rome [35, pp. 130–159] predicts that the intensification of the biomass production will be to the disadvantage of the environment and climate, and the IRENA forecast [36, pp. 45–58] agrees that many of the sustainability issues related to an intensive agriculture are not yet solved.

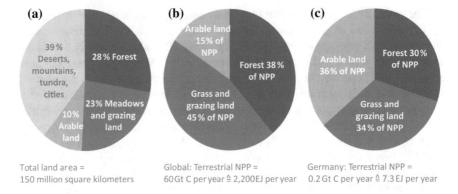

Fig. 4.6 Use of the terrestrial land surface and the respective proportions of net primary
production [2]. a Global distribution of terrestrial areas into forests, meadows and pastures, arable
land and areas that are unsuitable for forestry and agriculture uses. b Proportions that the different
types of land use areas contribute to the global net primary production (NPP). c Proportions that
the different types of land use areas contribute to the net primary production (NPP) in Germany

4.3.2 The Balance Sheet for Plant Biomass Production
in the Countryside

About one third of the global land area (150 million square kilometers) consist of
deserts, mountains, tundra, cities or other areas that are not suitable for forestry
and agriculture (Fig. 4.6a). The availability of biomass as a source of energy is
higher in the thinly populated countries such as the USA or Brazil with population
densities of 30 or, respectively, 20 people per square kilometer as compared to that in
Germany with a population density of 230 people per square kilometer [37]. The total
area under cultivation in the world currently supplies renewable plant biomass (net
primary production, see BOX XIV) annually with a weight of 60 Gigatonnes carbon
and an energy content of 2200 EJ. The carbon content of dried vegetable biomass
corresponds to about 50%. The average energy content (gross calorific value) of dried
biomass is 37 kJ/g of carbon.

About 38% of the global net primary production (NPP) is derived from forest
areas, 45% from pastures and grazing land, and 15% from arable land (Fig. 4.6b).
Over 50% of the global NPP is not available to mankind because it is either located
underground (30–40%) or in inaccessible locations (20%). People use about 10% of
the annual global renewable biomass, of which 7% is used for nutrition and animal
foods, 1% of industrial purposes, and 2% for energy generation.

The annual renewable plant biomass in Germany corresponds to a carbon content of about 0.2 Gigatonnes with an energy content of 7.3 EJ (Fig. 4.6c). About 30% of the German NPP) is from woodlands, 34% from pastures and grazing land, and 36% from arable land. About 0.09 Gigatonnes carbon with an energy content of 3.4 EJ are either harvested by mankind or grazed by animals annually from the NPP in Germany. This corresponds to 24% of the German primary energy consumption of 14 EJ. For example, in 2010 Germany received through imports an additional 0.07 Gigatonnes carbon from foreign NPP. This figure includes the products manufactured from biomass that were produced in foreign countries but were consumed in Germany.

BOX XIV: Net primary production of plant biomass

The net primary production (NPP) describes the amount of plant biomass that grows in one year within a specified area. The emissions from volatile organic compounds and excretions from the roots as well as surface and subsurface leaf litter that is decomposed within one year are neglected. The NPP is dependent on the sun radiation, the length of the vegetation growth period, the temperature and the water and mineral supply (Fig. 4.7).

The global net primary production has decreased by about ten percent over the past 200 years, despite an increase in the use of fertilizers, plant breeding and irrigation. This is mainly because the large areas of land used for agriculture have a lower NPP as the respective land with original natural vegetation. Furthermore, the NPP has declined because of the sealing of land and the deterioration in the quality of the soil. The decline in the NPP in many areas could not be compensated by an increase in the NPP from crops by fertilizing, irrigation and cultivation in some of the more intensively farmed areas such as in northwest Europe or the Nile valley. A further net decline in the global NPP is to be expected in the future as a result of soil erosion, salinization of the soils, and excessive construction of buildings rather than an increase from the expansion of the agrarian areas, use of fertilizers, irrigation and plant cultivation [34, 38–41].[16]

The continual increase of the harvest yields that has been achieved during the past fifty years has slowed down in recent years, and was mainly due to the redistribution of the NPP within plants (for example, more cereal grains with lower plant growth) and only insignificantly to the increase in the NPP. The earlier rates of increase of the harvest yields were not achieved by plant cultivation only but also by using fertilizers and by improved irrigation. It is questionable if, in future, it will be possible to further increase the NPP above its natural potential in larger areas over longer time periods [2, p. 16].

---

[16]See also Leopoldina [2, p. 16].

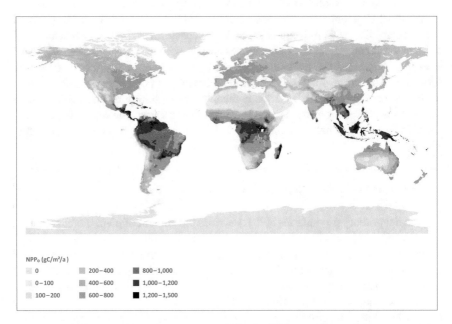

Fig. 4.7 Global terrestrial net primary production in 2000 (Haberl et al. [63], accompanying information material, see SI Fig. 2). The global terrestrial area is about 150 million square kilometers, of which only 100 million square kilometers is covered by vegetation. The average NPP per square kilometer is about 430 grams carbon per year ($gC/m^2/a$) relative to the total land area, and about 650 $gC/m^2/a$ relative to the vegetated areas. (*Source* Haberl et al. [63], with permission of the PNAS)

4.3.3 Biomass from Forests and Woodlands

About 41.6 million square kilometers, or about 28%, of the global land surface is covered by woodlands and forests (Fig. 4.6a). For comparison, an average of 32% of the land surface area in Germany is covered by woodlands and forests [42]. The global net primary production of the woodlands and forests is about 23 Gigatonnes carbon per year. By using the carbon balance, it can be shown that a harvest index[17] of 12–13% should not be exceeded for a sustainable exploitation of the forests with respect to the conservation of the forest and a climate-neutral utilization. The sustainability issue must be considered over a reference period of decades [2, Appendix 1]. The amount of timber that can be sustainably harvested each year is estimated

---

[17]The harvest index is derived from the percentage proportion of the total net primary production (on the surface and subsurface) that is sustainably harvestable as wood. With respect to the definition of a harvest index for wood it must be taken into consideration that most of the NPP in these long-lived flora is used for the continuous regrowth of leaves and roots. This NPP cannot be harvested as wood, which reduces the harvest index for trees.

to be about 3 Gigatonnes carbon, which corresponds to a calorific value of about 110 EJ. Even if all the timber that can be sustainably harvested is used for energy generation, then the proportion contributed by wood to the current global energy consumption (560 EJ/year) is only about twenty percent. This assumption is even then unrealistic because timber is also used for other purposes such as in construction or in the manufacturing of paper. An intensive use of timber for energy generation represents a serious risk that the integrity and sustainability of the forests and woodlands would be sacrificed for an increase in energy generation without any saving for climate change [43, 44]. According to information from the Food and Agricultural Organization of the United Nations (FAO), an average of about six percent of the global primary energy consumption in 2011 was provided by burning of wood,[18] which corresponds to a calorific value of about 30 EJ. The proportion of wood in the total energy consumption varies regionally. For example, in Africa the average proportion of 27%, which is the highest in the world, and in North America it is 2%, which is the lowest [45, p. 32].

There are 11.4 million hectares of forest and woodland in Germany, with a timber resource of 336 cubic meters per hectare and an average timber growth of 11.2 cubic meters per hectare and year. In the last 10 years, the forested area has increased by 0.4%. The forest is on average 77 years old, and 43% of the trees are deciduous. The trend from pure conifer populations towards structurally enriched mixed forests in locations suited to the local conditions, and the limitations on logging is continuing and contributes to stabilizing the climate. An average of only 76 million cubic meters of raw timber (harvested cubic meter without bark) was used per year from 2002–2012 [42].

4.3.4 Biomass from the Agricultural Industry

Currently 49 million square kilometers (33%) of the global surface area (150 million square kilometers) are being worked for agriculture (Fig. 4.6a), of which about 34 million square kilometers (23%) are grasslands and pastures with a global NPP of 27 gigatonnes carbon per year, and about 15.3 million square kilometers (10%) are arable fields with a global NPP of 10 gigatonnes carbon per year. About six gigatonnes of carbon, out of the total NPP derived from these areas (37 gigatonnes carbon per year), are currently used for the nutrition of humans and farmed animals as well as for the supply to mankind with materials (such as cotton) and energy (0.72 gigatonnes carbon per year, or 25 EJ/year). The remaining 31 gigatonnes carbon, which are mostly unsuitable for human consumption as they are subsurface roots (30–40%) or indigestible (lignocellulose), must be divided among all other living creatures.

More than one billion people are currently suffering from under-nourishment and inadequate supply of nutrients because of regional shortages of foodstuffs. Another billion people suffer from insufficient amounts of vitamins and/or trace elements

---

[18] About 55% of the harvested timber was used for meeting the primary energy consumption.

(hidden hunger). In future practically all of the agrarian biomass on Earth that can be sustainably harvested will be required to address this situation as nutrition for people (cereals, maize, rice, vegetables, salads, fats and oils), as animal feed, for materials (for example as cotton), or as the basis material for platform chemicals. Practically all of the agrarian biomass will be required in 2050 for the forecasted global population of nine to ten billion people. As a result, only small amounts of the agricultural biomass will be available for energy generation and only the waste from agrarian biomass will remain for use in energy generation, although these quantities need not be insignificant.

For example, 18 megatonnes of carbon (0.64 EJ) is generated annually from agrarian waste in Germany. Straw (20 megatonnes carbon; 0.7 EJ) is also in part available as a source of energy although it must be noted that most of the straw must be ploughed back into the soil to improve fertility [2].

Undeveloped potential

It is legally forbidden to dump biomass waste in Germany, and therefore facilities have already been established to process this material (for example, in biogas plants). In the rest of the world, however, most of the biomass waste is dumped whereby, due to the natural composting processes, large quantities of greenhouse gases especially methane are emitted. As an example, the manufacturing of one tonne of palm oil results in 4.4 tonnes of solid and liquid waste materials that, if dumped, can produce about 1.2 tonnes CO_2-equivalent greenhouse gasses [46]. The use of such waste for energy generation could reduce these greenhouse gas emissions.

The estimates of how much bioenergy derived from agrarian biomass will be available in 2050 are widely different. They range from 50 EJ/year (the current status) to 500 EJ/year.[19] The differences can be explained by, among others, the different assumptions made for the various parameters (see Sect. 4.3.1).

The production of animal products, such as meat and milk, requires significant amounts of vegetable biomass that are dependent on the type of animal and its feeding requirements. The different animal species convert feedstuff into meat with differing degrees of efficiency. For example, the production of one kilogram of beef meat (red meat) requires under certain circumstances about 13 kg cereal and 30 kg hay, whereas the production of one kilogram of chicken meat (white meat) requires only about 2.3 kg grain feed. About 40% of the total global production of cereal grain is used on average for animal feed to produce meat, milk, cheese and other foodstuffs derived from animals. This trend is increasing. Today about 75% of the area used by the global agriculture industry is required for raising farm animals.

About two thirds of the total nutritional energy in Germany and most of the other European countries, is derived from vegetable products and about one third from animal products (meat, milk, eggs, etc.). A vegetarian diet could release up to fifty percent of the agrarian areas for bioenergy [50]. The production chains for foodstuffs are currently very inefficient. About sixty percent of the harvest are lost due to shortfalls in the agriculture and the supply chains as well as throw-away

---

[19]Fritz et al. [47]; Recommendations of the BÖR [48]; IPCC [49, Chap. 2]: Leopoldina [2, p. 16].

disposal from supermarkets and households. A sustainable agricultural industry could provide a substantial reduction in these losses [2, p. 33], and would also reduce the greenhouse gas emissions, quantity of nitrates in the groundwater, and other environmental impacts.

4.3.5 Lignocellulose

It is often noted that a larger percentage of the net primary production would be available for energy generation if the lignocellulose in field plants and grasses could be converted to bioethanol (the so-called second generation fuels).[20] Because cellulose is indigestible by humans, this process would not lead to a competition between fuels and nutrition. However, this argument does not consider that ruminant animals (cattle, sheep, goats, etc.), which are an important source for animal products in foodstuff for humans, can live mainly on cellulose. Furthermore, the lignocellulose content in field plants is also needed to regenerate the carbon content in the soils. Cellulose and wood are above all plant products with many possible uses, such as in the construction and paper industries.

4.3.6 Land, Soils, Water, Nitrogen, Phosphorus, and Potassium

There are limits to the use of land, soils, fresh water, nitrogen and phosphorus by mankind. Excessive use can lead to inversions in the environmental conditions (for example, eutrophication of waters[21]). In this context, an international team of scientists has defined nine essential ecological parameters that are massively, and in part irreversibly, changing due to excessive use and overload—these include water usage, the global nitrogen cycle and biodiversity (decline in the variety of species). These ecological parameters are described as "planetary boundaries". The scientists assume that the planetary boundaries are already exceeded today for biodiversity, the global phosphorus and nitrogen cycles, climate change and use of land [52, 53].

Changes in land use

The areas (grasslands, pastures and arable land) used by the global agricultural industry increased by 1.54 million square kilometers between 1985 and 2000. In particular,

---

[20]Peplow [51].

[21]Eutroph means rich in nutrients. In this context it refers to an anthropogenic over-fertilization of water that can result in a disproportional growth of vegetation and therefore extensive imbalances in their environment (lack of oxygen, drying up of land, fauna extinction, excessive growth of specific plant species, etc.).

the agricultural areas have increased significantly in the tropics, whereas in the temperate zones the area required for agriculture has remained constant or even declined. The result is a net redistribution of the agricultural areas towards the tropics [2, p. 14].

A further expansion of the areas used for agriculture is problematic because globally 70% of the meadows, 50% of the savanna, 45% of the deciduous woodlands in the temperate zones, and 27% of the tropical forest areas have today either already been occupied or cleared. Mankind has already acquired[22] for agriculture and forestry more than 40% of the total global regrowth of biomass that must be shared by all living creatures. Nonetheless, it is expected that the areas intensively used for agricultural production will expand by about five percent between 2005 and 2050, and this includes all the related risks. The change of land use for agricultural purposes has immense impacts on the habitats, biological diversity, carbon storage, soil conditions, availability of drinking water, and greenhouse gas emissions (N_2O, CH_4 and CO_2). The intensive agricultural practices, together with the changes in land use caused by agriculture, contribute globally about twenty percent to the increase in the concentration of greenhouse gases in the atmosphere [2, p. 28, 54, p. 4].

Soils

The stability of soils is essential for their fertility. Soil stability means that the soil growth rate is the same or greater than the rate of soil loss. Soils are developed when the solid rock is broken down and converted into new mineral components by physical, chemical and biological processes, supported by organisms in the soil. The particles that derive from this process are combined with the decomposed biomass and living microbes into larger aggregations. These aggregations of mineral and organic nutrients are then processed by the living microbes so that the minerals can be used by plants. The pores within and between the soil aggregates are important reservoirs for moisture. They are essential for the biological growth, facilitate the water run-off, and make it possible for oxygen to be supplied through the roots of the plants.

The soils together contain two to three-times more carbon than the atmosphere (820 gigatonnes carbon) or terrestrial vegetation (800 gigatonnes carbonate).[23] The carbon occurs in soils mainly as organic material that is often very stable and exists for thousands of years, but some is also easily broken down. The stability of organic substances in soils depends not only on the molecular structure of the organic material, but also on environmental factors such as temperature, water activity and presence of microorganisms.

In some parts of the world the soil is lost one hundred times faster than it is formed [2, p. 36f]. This loss of soils can occur in many ways: for example, because of soil

---

[22]"Acquired" describes the difference between the NPP of the potential vegetation, which is vegetation that would grow in a specific area under the prevailing local conditions such as soil and today's climate but without the presence of humans, and the proportion of the NPP of the currently dominant vegetation that remains in the eco-system after harvesting for mankind (see Leopoldina [2, pp. 116–132]).

[23]The carbon that is subject to the carbon cycle is balanced in this way. Carbon that is contained in rocks, such as limestones, is not taken into account.

erosion by rain, dust storms, deterioration by environmental pollution, by salt from evaporation of irrigation water, by compaction from heavy mechanical equipment, by soil carbon that is oxidized to CO_2, or by soils that are literally sealed when covered by urbanization such as roads and houses [55].

Water

Plants usually need a lot of water for an optimal growth. Because of the moderate temperatures and the relatively high precipitation, the cultivation of plants in Germany does not normally need irrigation, and where it is required there is normally sufficient water available. In countries with less precipitation as compared to Germany, high yields can only be achieved with the help of irrigation. However, soils can also become salty because of the irrigation. Irrigation is generally very important for agricultural productivity, and globally it is estimated that about 2800 cubic kilometers per year (70% of the global extraction of fresh water) are extracted from groundwater, lakes and rivers for this irrigation. This extracted water is used for irrigation of about 24% of the arable fields that produce about 34% of the agricultural products. The areas under irrigation in the world have about doubled during the past 50 years. At the same time, significant areas have been lost to agriculture because of salinization. The planetary boundary for global consumption of fresh water is estimated to be nearly 4000 cubic kilometers per year [52].

If the sea water is included, then there is an unlimited quantity of water on Earth. However, sea water can only be used if it is treated in a desalination plant, which requires a significant amount of energy. Because plants need large quantities of water, growth with desalinated water consumes up to fifty percent of the energy that can be released by combustion of the resultant biomass. If solar energy is used for desalination, then the energy consumption is not a problematic issue in itself, but the same solar energy could be more efficiently directly used for electricity generation, rather than for desalination.

Nitrogen fertilizers

Plants consist of carbon, to a proportion of about fifty percent, and of significantly less nitrogen (N) depending on the plant species. For example, spruce wood with bark has a nitrogen content of 0.13% and whole wheat plants a content of 1.4%. While the carbon content in different plant tissues is relatively constant, the nitrogen content varies from 0.03 to 7% depending on the tissue type. Nitrogen is an essential component of proteins and nucleic acids as well as chlorophyll. Atmospheric nitrogen is fixed with the help of micro-organisms,[24] and is available in virtually unlimited amounts. In order to achieve higher yields, it is necessary to use a nitrogen fertilizer (ammonium—NH_4^+, urea—$CO(NH_2)_2$ or nitrate—NO_3^-). One hectare arable land in Germany was fertilized with an average of 100 kg nitrogen in 2010 [56, 57].

Globally about 110 million tonnes nitrogen are introduced annually as chemically produced nitrogen fertilizers (Haber-Bosch-Process) into the natural nitrogen cycle.

---

[24]Nitrogen generally occurs as molecular N_2.

Additionally, up to 20 million tonnes nitrous oxide (NO_2) are derived from the air as a result of combustion processes. The anthropogenic introduction of nitrogen exceeds the quantity of nitrogen that is biologically fixed by micro-organisms (about 40 million tonnes nitrogen per year) [52]. The nitrogen cycle is more seriously influenced by fertilizing, with unforeseeable consequences, than the carbon cycle is affected by the combustion of fossil fuels.

The manufacturing of fertilizers by the Haber-Bosch process has increased by more than 800% during the past fifty years. The Haber-Bosch process is very energy-intensive and, despite improved efficiencies in the processing, it uses a little more than one percent of the global energy consumption. There is unlikely to be a shortage of nitrogen fertilizers because they can be manufactured chemically. More importantly, the effects on the environment, both as greenhouse gas emissions and the effects on the nitrogen cycle, by their manufacture as well as agricultural use, could be regarded as possible constraining factors in the future.

Phosphate fertilizers

Phosphate is also an important constituent of plants. It occurs in the nucleic acids DNA and RNA, in intermediates of metabolism such as glucose-6-phosphate and in coenzymes such as adenosine-triphosphate (ATP), which acts as the energy carrier within all living cells. Although there is often sufficient phosphate in the soils, this cannot always be absorbed by the plants because it occurs as insoluble calcium-phosphate that can only be slowly mobilized by plants. Certain microbes (mycorrhiza) interact with the roots of the plan and enable the intake of phosphate by excreting organic acids (for example, citric acid) that accelerate the mobilization process and, as a result, the plant growth is sustained. However, in order to achieve high yields, additional fertilizing with phosphates is usually necessary. In Germany one square kilometer of agricultural land was fertilized with about 14 kg phosphate[25] per year (excluding liquid manure) in 2010. Globally about 44 million tonnes phosphate is spread onto the fields, of which 80% end up as compounds that are either not, or only slowly, accessible for plants. About 3.7 million tonnes phosphate is used additionally for animal feed and four million tonnes for industrial processes [56, 57]. An estimated three million tonnes flows through the drainages into the sea.

The German agricultural industry currently uses about 650,000 tonnes phosphate per year. About half of this must be imported, and nearly 300,000 tonnes can be covered by liquid manure. Additionally, significant quantities of phosphate can be extracted from the urban and industrial wastewater treatment plants. However, until now, sewage sludge is burnt which results in insoluble phosphate complexes that require additional energy for their recovery and conversion to plant-fertilizing phosphate.

Furthermore, industrial and urban waste is normally contaminated with heavy metals, and it is therefore generally forbidden to be used as fertilizer in forests and on agricultural fields. However, although it is technically possible to remove the contaminant metals, but only with additional energy costs.

---

[25]Calculated in the compound P_2O_5.

The phosphate content of biomass derived from producing biogas from agricultural plants can be completely recovered, if the residues from the fermentation process are used for fertilizing. During the microbial production of biogas, micro-organisms are enriched in phosphate. Most of the phosphate in the residues from the fermentation process is biologically available for the growth of plants. Although the global occurrences of phosphates are finite, phosphate will not be a constraining factor for the agricultural industry in the foreseeable future to mine production is currently estimated to be about 300 years [58: Phosphate Rock, p. 118ff]. This early warning indicator does not suggest any shortages of supply for phosphate (compare Sects. 3.1 and 5.4.2) [6].

Potassium fertilizers

Potassium is an important component of all living cells, including those plants in which high concentrations occur and, for example, is required for the maintenance of the osmotic (Turgor) pressure in the plant cells, especially in leaves. An average of 22-kg potassium[26] were added as fertilizer to each hectare of agricultural land in Germany in 2010. Most potassium salts are easily soluble and are washed by rain out of the soils, and therefore it is necessary to regularly apply fertilizer with potassium salts in order to ensure a good plant growth [56, 57].

All the known geological deposits in the world contain an estimated 210 billion tonnes K_2O, of which up to 16 billion tonnes K_2O can be exploited with the currently available technology. There is an almost unlimited quantity of potassium in sea water that can be concentrated by water evaporation and recovered as K_2O, examples of which are the operations on the Dead Sea in Israel and Jordan as well as the Qarhan Lake in China. A shortage of potassium as a resource is therefore very unlikely.

4.3.7 *Economic Aspects of the Utilization of Biomass*

The cultivation and harvesting of biomass obviously requires a certain energy input that must be compared to the useful energy output. The energy return on investment (EROI, energy output divided by the energy input) for firewood is about ten, for bioethanol derived from maize (USA) or electricity from bioethanol derived from maize it is only about 1.5 [2, p. 22ff].

Compared to other energy technologies such as photovoltaic or wind power facilities, bioenergy only has a low efficiency for the area used of less than 0.5 W/m^2. The efficiency for photovoltaic modules in Germany is generally more than 5 W/m^2, and that for wind power facilities in the countryside it is between 2 and 3 W/m^2 (average efficiency per year: 365 days at 24 h). This higher efficiency is also reflected by higher EROI values, seven for photovoltaic and up to 18 for wind power facilities. Furthermore, after they have been installed in the countryside there are only low additional maintenance costs for photovoltaic cells and wind power facilities.

---

[26]Data for potassium monoxide (K_2O) as a reference for the potassium content.

Except for the biomass that is derived from sustainably managed woodlands and forests, bioenergy contributes the least to reducing greenhouse gas emissions and pays the highest financial price per tonne of saved-CO_2 as compared to all other solar-based energy systems [59]. Bioenergy also shows a lower efficiency for the area utilized as compared to other renewable energy systems such as wind and solar energy. The potential for bioenergy relative to the primary energy consumption is estimated to be low if all the sustainability criteria, the competition between the "tank and the plate", and the nutrition preferences of mankind are taken into account. The utilization of biomass as an energy source should primarily be restricted to wood derived from sustainably managed forests and woodlands and to agrarian waste. Based on this, the utilization of bioenergy is only meaningful if it brings the most advantages to the total system.

The advantage of biofuels (bioethanol, biodiesel, biogas) is that they have a high energy density allowing them to replace fossil fuels in transportation especially by airplanes, heavy duty trucks and cargo ships that will be difficult to run on batteries with a much lower energy density [60]. However, in densely populated regions such as Germany other efficiency measures for the transport sector should also be considered such as reducing traffic by changing the spatial planning and/or expanding the public transport facilities as well as enforcing speed limits. This could save more fossil fuels than can be sustainably produced from bioenergy sources [2]. The fact that bioenergy can be stored is another important advantage. Thus stored bioenergy can be used to generate electricity during longer periods where there is not enough wind and/or sunlight or used as a source of balancing energy.

Supply Status for Biomass

The proportion of bioenergy in the global energy consumption is currently ten percent and is expected to decrease rather than increase because of the rising global population and resultant increased competition for the use of land. This is especially the case if the biomass requirements are to be met by sustainable production.

The quantities of biomass that are produced in Germany are theoretically almost entirely required for non-energy uses. The waste from this biomass would essentially be the only material that could be used for energy generation, if Germany did not meet its demand for biomass from additional imports. The imported biomass results directly or indirectly to shortages of biomass for nutrition elsewhere, to the deforestation of rain forests, to the decline of biodiversity, and/or to the formation of greenhouse gases.

The use of lignocellulose, for example from wood and grasses, in energy utilities does not compete directly with human nutrition, and is therefore often regarded as to have a greater potential for energy generation. However, it must be considered that lignocellulose is required for animal feedstock, for soil regeneration as well as other uses.

Intensive agricultural practices together with the related changes to land use contributes globally to about 20% of the increase in the greenhouse gas concentration in the atmosphere. Only the use of wood from sustainably managed forests and woodlands is overall greenhouse gas neutral, but a more intensive use of this biomass would also lead to greater greenhouse gas emissions.

Intensive agricultural practices result in a decline in biodiversity, increase in water consumption and contamination of waters by excessive use of fertilizer, as well as soil degradation. In some parts of the world the soils are currently being lost one hundred times faster than they are being developed. Intensive irrigation in arid regions cannot only result in a loss of freshwater reservoirs but also to a salinization of agricultural areas. The utilization of biomass derived from intensive agricultural practices for energy generation is therefore not sustainable.

Currently about sixty percent of the global harvest is lost due to deficiencies in agriculture and in the supply chains, as well as to wasteful practices in supermarkets and households. A significant reduction in these losses would reduce the demand for biomass for nutrition, and alleviate the environmental impacts.

The three fertilizers, nitrogen, potassium and phosphate that are essential for plant growth, are available in sufficient quantities. An excessive use of fertilizers, as is required for intensive agricultural practices, affects the global materials cycle as discussed with, in part, unforeseeable consequences. It has been suggested that the planetary boundaries for limits to use have already been exceeded for biodiversity, the phosphorus and nitrogen cycles, climate change and land use.

Phosphate is the only fertilizer resource that is not available in unlimited quantities. Therefore, if biomass is to be sustainably utilized for energy generation, it is important to recover the phosphate in the biomass for use as fertilizer in agriculture. This is already practiced in, for example, biogas facilities.

As compared tos other renewable energy systems, bioenergy has the lowest efficiency for the area used. It also contributes less than wind and solar energy to the reduction of greenhouse gases, and the costs per tonne of saved CO_2 are higher. The higher energy density and the storability of bioenergy are advantages for its use in transport or for generating electricity during longer periods of calm winds and low solar radiation.

References

(Note: All Web links listed were active as of the access date but may no longer be available.)

1. Bringezu, S./Schütz, H./Arnold, K./Merten, F./Kabasci, S./Borelbach, P./Michels, C./Reinhardt, G. A./Rettenmaier, N.: "Global Implications of Biomass and Biofuel Use in Germany – Recent Trends and future Scenarios for domestic and foreign agricultural Land Use and resulting GHG Emissions". In: *Journal Cleaner Production*, 17, Elsevier Ltd., 2009, pp. 57–68.
2. Nationale Akademie der Wissenschaften Leopoldina: *Bioenergie – Möglichkeiten und Grenzen* (Stellungnahme der Leopoldina), Halle (Saale) 2013. URL: http://www.leopoldina.org/uploa ds/tx_leopublication/2013_06_Stellungnahme_Bioenergie_DE.pdf [accessed: 01.11.2014].
3. Bundesministerium für Umwelt, Naturschutz und Reaktorsicherheit/Bundesministerium für Ernährung, Landwirtschaft und Verbraucherschutz (Hrsg.): *Nationaler Biomasseaktionsplan für Deutschland – Beitrag der Biomasse für eine nachhaltige Energieversorgung*, Berlin 2010.
4. Vidal, O./Goffé, B./Arndt, N.: "Metals for a low-carbon Society" (Supplementary Information). In: *Nature Geoscience*, 6, 2013, pp. 894–896.
5. Hertwich, E.G./Gibojn, T./ Bouman, E. A./Arvesen, A./Suh, S./Heath, G. A./Bergesen, J. D./Ramirez, A./Vega, M. I./Shi, L.: "Integrated life-cycle Assessment of Electricity-Supply Scenarios confirms global environmental Benefit of low-carbon Technologies". In: *Proceedings of the National Academy of Sciences of the USA*, 2014. URL: http://www.pnas.org/conte nt/suppl/2014/10/02/1312753111.DCSupplemental [accessed: 28.10.2014].
6. Scholz, R./Wellmer, F.-W.: "Approaching a dynamic View on the Availability of Mineral Resources: What we may learn from the Case of Phosphorus?". In: *Global Environmental Change* 23: 1, 2013, pp. 11–27.
7. Hiller, K.: "Explorations-Förderprogramm DEMINEX (1969–1989)". In: *Geologisches Jahrbuch Reihe A*, Heft 127, 1991, p. 289–298.
8. Anger, G.: "Deutscher Auslandsbergbau – unternehmerische Aktivitäten und verbandliche Gemeinschaftsaufgaben". In: *Jahrbuch Bergbau, Öl und Gas, Elektrizität, Chemie*, 1990/91, pp. 1–26.
9. Wellmer, F.-W.: "The Concept of Lead Time". In: *Minerals Industry International*, 1005, 1992, pp. 39–40.
10. Kingsnorth, D. J.: *The global Rare Earths Industry: the Supply Chain Challenges* (Company paper IMCOA [Industrial Minerals Company of Australia Pty. Ltd.]), 2012.
11. Schodde, R. C.: "Recent Trends in Copper Exploration – are we finding enough?". In: *International Geological Congress IGC Brisbane*, Australien, 05. – 10. August 2012), 2012. URL: http://www.minexconsulting.com/publications/IGC%20Presentation%20Aug%2 02012%20PUBLIC.pdf [accessed: 15.01.2015].
12. Schodde, R.C.: "Global Outlook and Development Trends for Copper". In: *Philippines Mining Conference*, Manila, 20.09.2012. URL: http://www.minexconsulting.com/publications/Coppe r%20Outlook%20-%20PMC%20presentation%20Sept%202012.pdf [accessed: 13.03.2014].
13. Matthews, L.: *Speaking Notes for Lloyd Matthews, Minister of Mines and Energy*. Update on the Voisey's Bay Negotiations to St. John's Board of Trade, 11. October 2001. URL: http://www.r eleases.gov.nl.ca/releases/speeches/2001/outlook2001/BoardOfTradeOct2001.htm [accessed: 12.08.2014].
14. Bundesanstalt für Geowissenschaften und Rohstoffe: *Energiestudie 2014: Reserven, Ressourcen und Verfügbarkeit von Energierohstoffen* (18), Hannover 2014.
15. International Energy Agency: *World Energy Outlook 2014*, Paris: IEA 2013. URL: http://www. worldenergyoutlook.org/publications/weo-2013/ [accessed: 17.11.2014].
16. Organization of the Petroleum Exporting Countries: *World Oil Outlook 2014*, Wien 2014. URL: http://www.opec.org/opec_web/static_files_project/media/downloads/publications/WO O_2014.pdf [accessed: 17.11.2014].

17. British Petroleum: *Statistical Review of World Energy*, London 2015. URL: http://www.bp.co m/content/dam/bp/pdf/Energy-economics/statistical-review-2015/bp-statistical-review-of-wo rld-energy-2015-full-report.pdf [accessed: 28.08.2015].
18. Arbeitsgemeinschaft Energiebilanzen e.V.: *Energieverbrauch in Deutschland im Jahr 2014*, Berlin 2015. URL: http://www.ag-energiebilanzen.de/ [accessed: 23.03.2015].
19. US Environmental Protection Agency: *EPA's Study of Hydraulic Fracturing for Oil and Gas and its Potential Impact on Drinking Water Resources,* 2015. URL: http://www2.epa.gov/hfst udy [accessed: 20.02.2015].
20. Bundesanstalt für Geowissenschaften und Rohstoffe: *Abschätzung des Erdgaspotenzials aus dichten Tongesteinen (Schiefergas) in Deutschland*, Hannover: Bundesanstalt für Geowissenschaften und Rohstoffe 2012.
21. Bundesanstalt für Geowissenschaften und Rohstoffe: *Schieferöl und Schiefergas in Deutschland – Potenziale und Umweltaspekte,* Hannover 2016.
22. Ewen, C./Borchardt, D./Richter, S./Hammerbacher, R.: *Risikostudie Fracking – Sicherheit und Umweltverträglichkeit der Fracking-Technologie für die Erdgasgewinnung aus unkonventionellen Quellen* (Summary Report), Darmstadt 2012.
23. Meiners, H.G./Denneborg, M./Müller, F./Bergmann, A./Weber, F.-A./Dopp, E./Hansen, C./Schüth, C./Buchholz, G./Gaßner, H./Sass, I./Homuth, S./Priebs, R.: *Umweltauswirkungen von Fracking bei der Aufsuchung und Gewinnung von Erdgas aus unkonvetionellen Lagerstätten – Risikobewertung, Handlungsempfehlungen und Evaluierung bestehender rechtlicher Regelungen und Verwaltungsstrukturen* (Study for the Umweltbundesamt), Aachen, Mühlheim a. d. Ruhr, Berlin, Darmstadt: ahu AG Wasser Boden Geomatik/IWW Rheinisch-Westfälisches Institut für Wasser – Beratungs- und Entwicklungsgesellschaft mbH/Gaßner, Groth, Siederer & Coll. Rechtsanwälte Partnerschaftsgesellschaft/Technische Universität Darmstadt, Institut für Angewandte Geowissenschaften, Fachgebiet Angewandte Geothermie 2012.
24. Ministerium für Klimaschutz, Umwelt, Landwirtschaft, Natur- und Verbraucherschutz des Landes Nordrhein-Westfalen: *Fracking in unkonventionellen Erdgas-Lagerstätten in NRW*, Summary report about the expert opinion: „Gutachten mit Risikostudie zur Exploration und Gewinnung von Erdgas aus unkonventionellen Lagerstätten in Nordrhein-Westfalen (NRW) und deren Auswirkungen auf den Naturhaushalt, insbesondere die öffentliche Trinkwasserversorgung", Düsseldorf 2012.
25. Deutsche Akademie der Technikwissenschaften (acatech): *Hydraulic Fracturing – Eine Technologie in der Diskussion* (acatech Position), 2015. URL: http://www.acatech.de/de/projek te/laufende-projekte/hydraulic-fracturing-eine-technologie-in-der-diskussion.html [accessed: 01.10.2015].
26. Bundesanstalt für Geowissenschaften und Rohstoffe: *Deutschland – Rohstoffsituation 2013,* Hannover 2014.
27. Görner, K./Sauer D. U.: *Konventionelle Kraftwerke. Technologiesteckbrief zur Analyse "Flexibilitätskonzepte für die Stromversorgung 2050"*, München 2016.
28. Elsner, P./Fischedick, M./Sauer, D. U. (Hrsg.): *Flexibilitätskonzepte für die Stromversorgung 2050: Technologien – Szenarien – Systemzusammenhänge* (Schriftenreihe Energiesysteme der Zukunft), München 2015.
29. Nationale Akademie der Wissenschaften Leopoldina: *Bioenergy – Chances and Limits* (Stellungnahme der Leopoldina), Halle (Saale) 2012. URL: http://www.leopoldina.org/uplo ads/tx_leopublication/201207_Stellungnahme_Bioenergie_LAY_en_final_01.pdf [accessed: 18.02.2015].
30. Umweltbundesamt: *Emissionen aus der Landwirtschaft im Jahr 2010*, Dessau-Roßlau 2014. URL: http://www.umweltbundesamt.de/daten/land-forstwirtschaft/landwirtschaft/beitrag-der-landwirtschaft-zu-den-treibhausgas [accessed: 28.10.2014].
31. European Academies Science Advisory Council: *Policy Report 19 – The current Status of Biofuels in the European Union, their environmental Impacts and future Prospects,* 2012. URL: http://www.easac.eu/home/reports-and-statements.html [accessed: 28.10.2014].
32. Seppelt, R./Mahceur, A. M./Liu, J./Fehichel, E. P./Klotz, S.: "Synchronized peak-rate Years of global Resources Use". In: *Ecology and Society*, 19: 4, 2014, Article 50.

33. Bundesministerium für Ernährung und Landwirtschaft: *Statistischer Monatsbericht 03/2015 – Daten und Tabellen*, Berlin 2015. URL: http://www.bmelv-statistik.de/index.php?i d=139&ab=66 [accessed: 10.04.2015].

34. Haberl, H./Erb, K.H./Krausmann, F./Running, S./Searchinger, T.D./Smith, W.K.: "Bioenergy, how much can we expect for 2050?". In: *Environmental Research Letters*, 8, Article- Nr. 031004, 2013.

35. Randers, J.: *2052 – A global Forecast for the next forty Years,* A Report to the Club of Rome commemorating the 40th Anniversary of the "The Limits to Gowth", White River Junction, Vermont: Chelsea Green Publishing 2012.

36. International Renewable Energy Agency: *Global Bioenergy – Supply and Demand Projections, a Working Paper for REmap 2030*, Abu Dhabi, Bonn: IRENA 2014.

37. US Department of Energy: *US Billion-Ton Update: Biomass Supply for Bioenergy and Bio-products Industry*, Energy Efficiency and Renewable Energy Office of the Biomass Program 2011. URL: http://bioenergykdf.net / http://www1.eere.energy.gov/bioenergy/pdfs/billion_to n_update.pdf [accessed: 28.10.2014].

38. Zika, M./Erb, K. H.: "The global loss of net primary production resulting from human-induced soil degradation in dry lands". In: *Ecological Economics*, 69, 2009, pp. 3010–3018.

39. Zhao, M./Running, S. D. W.: "Drought-induced reduction in global terrestrial net primary production from 2000 through 2009". In: *Science*, 329, 2010, pp. 940–943.

40. Krausmann, F./Erb, K.-H./Gingrich, S./Hbaerl, H./Bondeau, A./Gaube, V./Lauk, C./Plutzar, C./Searchinger, T. D.: "Global Human Approriation of net primary Production doubled in the 20th Century". In*: Proceedings of the National Academy of Sciences of the USA*, 110, 2013, pp. 10324–10329.

41. Hejazi, M./Edmonds, J./Clarke, L./Kyle, P./Davies, E./Chaturvedi, V./Wise, M./Patel, P./Eom, J./Calvin, K./Moss, R./Kim, S.: "Long-term global Water Projections using six socioeconomic Scenarios in an integrated Assessment Modeling Framework". In: *Technological Forecasting and Social Change*, 81, 2014, p. 205–226.

42. Bundesministerium für Ernährung und Landwirtschaft (Hrsg.): *Der Wald in Deutschland – Ausgewählte Ergebnisse der dritten Bundeswaldinventur*, Berlin 2014. URL: http://www.bm el.de/SharedDocs/Downloads/Broschueren/Bundeswaldinventur3.pdf?__blob=publicationFi le [accessed: 13.03.2015].

43. Schulze, E. D./Körner, C. I./Law, B. E./Haberl, H./Luyssaert, S.: "Large-scale Bioenergy from additional harvest of Forest Biomass is neither sustainable nor Greenhouse Gas neutral". In: *Global Change Biology Bioenergy*, 4: 6, 2012, pp. 611–616.

44. Stephenson, N. L./Das, A. J./Condit, R./Russo, S. E./Baker, P. J./Beckman, N. G./Coomes, D. A./Lines, E. R./Morris, W. K./Rüger, N./Álvarez, E./Blundo, C./Bunyavejchewin, S./Chuyong, G./Davies, S. J./Duque, Á./Ewango, C. N./Flores, O./Franklin, J. F./Grau, H. R./Hao, Z./Harmon, M. E./Hubbell, S. P./Kenfack, D./Lin, Y./Makana, J.-R./Malizia, A./Malizia, L. R./Pabst, R. J./Pongpattananurak, N./Su, S.-H./Sun, I.-F./Tan, S./Thomas, D./van Mantgem, P. J./Wang, X./Wiser, S. K./Zavala, M. A.: "Rate of Tree Carbon Accumulation increases continuously with Tree Size". In: *Nature*, 507, 2014, pp. 90–93.

45. Food and Agriculture Organization of the United Nations: *Status of the World's Forests 2014 – Enhancing the socioeconomic Benefits from Forests*, Rom 2014. URL: http://www.fa o.org/3/a-i3710e.pdf [accessed: 28.09.2015].

46. Stichnothe, H./Schuchardt, F.: "Comparison of different Treatment Options for Palm Oil Production Waste on a Life Cycle Basis". In: *International Journal of Life Cycle Assessment*, 15, 2010, pp. 907–915.

47. Fritz, S./See, L./van der Velde, M./Nalepa, R. A./Perger, C./Schill, C./McCallum, I./Schepaschenko, D./Kraxner, F./Cai, X./Zhang, X./Ortner, S./Hazarika, R./Cipriani, A./di Bella, C./Rabia, A. H./Garcia, A./Vakolyuk, M./Singha, K./Beget, M.E./Erasmi, S./Albrecht, F./Shaw, B./Obersteiner, M.: "Downgrading recent Estimates of Land available for Biofuel Production". In: *Environmental Science & Technology* 47, 2013, p. 1688–1694.

48. BioÖkonomieRat: *Empfehlungen 3 – Nachhaltige Nutzung von Bioenergie*, Berlin: Forschungs- und Technologierat Bioökonomie 2012. URL: http://www.biooekonomierat.de/fileadmin/ templates/publikationen/empfehlungen/BioOEkonmieRat-Empfehlungen-Bioenergie.pdf [accessed: 29.10.2014].

49. Intergovernmental Panel on Climate Change: *Renewable Energy Sources and Climate Change Mitigation* (SRREN – Special report of the IPCC, Working Group III "Mitigation of Climate Change", Chapter 2 (Bioenergy), New York, NY: Cambridge Univ. Press 2011, pp. 214–331. URL: http://srren.ipcc-wg3.de/report [accessed: 28.10.2014].

50. Eisler, M.C./Lee, M.R.F./Tarlton, J.F./Martin, G.B./Beddington, J./Dungait, J.A.J./Greathead, H./Liu, J./Mathew, S./Miller, H./Misselbrook, T./Murray, P./Vinod, V.K./van Saun, R./Winter, M.: "Steps to sustainable livestock". In: *Nature*, 507, 2014, pp. 32–34.

51. Peplow, M.: "Cellulosic Ethanol Fights for Life". In: *Nature*, 507, 2014, pp. 152–156.

52. Rockström, J./Steffen, W./Noone, K./Persson, Å./Chapin III, F.S./Lambin, E./Lenton, T.M./Scheffer, M./Folke, C./Schellnhuber, H. J./Nykvist, B./de Wit, C. A./Hughes, T./ van der Leeuw, S./Rodhe, H./Sörlin, S./Snyder, P. K./Costanza, R./Svedin, U./Falkenmark, M./Karlberg, L./Corell, R.W./Fabry, V.J./Hansen, J./Walker, B./Liverman, D./Richardson, K./Crutzen, P./Foley, J.: "Planetary Boundaries: Exploring the safe operating Space for Humanity". In: *Ecology and Society*, 14(2): 32, 2009. URL: http://www.ecologyandsociety.o rg/vol14/iss2/art32/ [accessed: 28.10.2014].

53. Steffen, W./Richardson, K./Rockström, J./Cornell, S.E./Fetzer, I./Bennett, E.M./Biggs, R./Carpenter, S.R./de Vries, W./de Wit, C.A./Gerten, D./Heinke, J./Mace, G.M./Persson, L.M./Ramanathan, V./Reyers, B./Sörlin, S.: "Planetary Boundaries: Guiding human Development on a changing Planet". In: *Science*, 347: 6223, 2015.

54. Wegener, J./Theuvsen, L.: *Handlungsempfehlungen zur Minderung von stickstoffbedingten Treibhausgasemissionen in der Landwirtschaft*, Berlin: WWF Deutschland 2010. URL: http:// www.uni-goettingen.de/de/document/download/9fd9831506d1021458b96370775b432e.pdf/ 100720_Stickstoffbroschuere.pdf [accessed: 05.11.2014].

55. Banwart, S.: "Save our Soils". In: *Nature*, 474, 2011, pp. 151–152.

56. Bundesministerium für Ernährung und Landwirtschaft (Hrsg.): *Statistisches Jahrbuch über Ernährung, Landwirtschaft und Forsten 2013*, Münster: Landwirtschaftsverlag GmbH 2013. URL: www.bmelv-statistik.de; http://www.bmelv-statistik.de/fileadmin/sites/010_Jahrbuch/S tat_Jahrbuch_2013.pdf [accessed: 27.01.2015].

57. Food and Agriculture Organization of the United Nation: *Current world Fertilizer Trends and Outlook 2016*, Rom 2012. ftp://ftp.fao.org/ag/agp/docs/cwfto16.pdf [accessed: 28.10.2014].

58. US Geological Survey: *Mineral Commodity Summaries 2015*, Washington DC 2014. URL: http://minerals.usgs.gov/minerals/pubs/mcs/2015/mcs2015.pdf [accessed: 07.08.201].

59. Organisation for Economic Co-operation and Development: *Biofuel support Policies: An economic Assessment*, 2008.

60. Bley, T. (Hrsg.): *Biotechnologische Energieumwandlungen: Gegenwärtige Situation, Chancen und künftiger Forschungsbedarf* (acatech Diskutiert), Berlin, Heidelberg: Springer Verlag 2009.

61. Wellmer, F.-W./Dalheimer, M.: "The Feedback Control Cycle as Regulator of past and future Mineral Supply". In: *Mineralium Deposita*, 47: 7, 2012, pp. 713–729.

62. International Energy Agency: *Key World Energy Statistics*, Paris: IEA 2014.

63. Haberl, H./Erb, K.-H./Krausmann, F./Gaube, V./Bondeau, A./Plutzar, C./Gingrich, S./Lucht, W./Fischer-Kowalski, M.: "Quantifying and Mapping the global human Appropriation of net primary Production in Earth's terrestrial Ecosystem". In: *Proceedings of the National Academy of Sciences of the USA*, 104: 31, 2007, p. 12942–12947. URL: http://www.pnas.or g/content/suppl/2007/07/09/0704243104.DC1 [accessed: 12.02.2015].

64. Barthel, F./Busch, K./Könnecker, K./Thoste, V./Wagner, H.: "Zwanzig Jahre Explorations-förderung für mineralische Rohstoffe". In: *Geologisches Jahrbuch* Series *A*, Volume 127, 1991, pp. 271–288.

65. Keitel, H.-P.: *Rohstoffsicherheit für Deutschland und Europa*. In: Key note speech *3. BDI-Raw Materials Congress*, Berlin, 26.10.2010.

66. Wedig, M.: "Die Entwicklung des Auslandsbergbaus am Beispiel von FAB-Strategien". In: *Bergbau*, 1, 2014, pp. 4–9.
67. Bundesministerium für Bildung und Forschung: *Forschung für nachhaltige Entwicklungen (FONA) – Rahmenprogramm des BMBF*, Bonn, Berlin: BMBF 2009, p. 59. URL: http://www.fo na.de/mediathek/pdf/forschung_nachhaltige_entwicklungen_neu.pdf [accessed: 29.10.2014].
68. Bundesministerium für Bildung und Forschung: *Ressourceneffizienz potenzieren. Broschüre zum Förderschwerpunkt "Innovative Technologien für Ressourceneffizienz – rohstoffintensive Produktionsprozesse"* (r^2), Karlsruhe: Fraunhofer-Institut für System- und Innovationsforschung ISI 2010. URL: http://www.r-zwei-innovation.de/_media/r2_broschuere_web.pdf [accessed: 29.10.2014].
69. Bundesministerium für Bildung und Forschung: r^3 – *Strategische Metalle und Mineralien – Innovative Technologien für Ressourceneffizienz*, Bonn: BMBF 2013. URL: http://www.fona. de/mediathek/r3/pdf/131126_r3_Broschuere_barrierefrei.pdf [Stand: 29.10.2014].
70. Bundesministerium für Bildung und Forschung: *Wirtschaftsstrategische Rohstoffe für den Hightech-Standort Deutschland* (r^4), Bonn: BMBF 2012.
71. Bundesministerium für Wirtschaft und Energie: *Rohstoffstrategie der Bundesregierung – Sicherung einer nachhaltigen Rohstoffversorgung Deutschlands mit nicht-energetischen mineralischen Rohstoffen*, Berlin: BMWi 2010.
72. Bundesministerium für Umwelt, Naturschutz, Bau und Reaktorsicherheit: *Deutsches Ressourceneffizienzprogramm (ProgRess) – Programm zur nachhaltigen Nutzung und zum Schutz der natürlichen Ressourcen*, Berlin: BMUB 2012. URL: http://www.bmub.bund. de/fileadmin/Daten_BMU/Pools/Broschueren/progress_dt_bf.pdf [accessed: 29.10.2014].
73. Bundesministerium für Wirtschaft und Energie: *Bekanntmachung im Rahmen der Rohstoffstrategie der Bundesregierung: Richtlinien über die Gewährung von bedingt rückzahlbaren Zuwendungen zur Verbesserung der Versorgung der Bundesrepublik Deutschland mit kritischen Rohstoffen (Explorationsförderrichtlinien)*, Berlin: BMWi 2012.
74. Le Parisien: *"Arnaud Montebourg: La renaissance d'une compagnie nationale des mines,"* 2014. URL: http://www.leparisien.fr/economie/arnaud-montebourg-la-renaissance-d-une-co mpagnie-nationale-des-mines-21-02-2014-3611305.php [accessed: 01.10.2015].
75. Frankfurter Allgemeine Zeitung: *Frankreich gründet einen Staatsbergbaukonzern*, 21.02.2014. URL: http://www.faz.net/aktuell/wirtschaft/wirtschaftspolitik/industrie-politik-frankreich-gru endet-einen-staatsbergbaukonzern-12814521.html [accessed: 01.10.2015].

Chapter 5
The Raw Material Requirements for Energy Systems

The new energy systems will be significantly more diverse as compared to the traditional energy technologies. Increasingly distinctive and mostly decentralized technologies will be added to the traditional generating technologies, such as coal and natural gas power stations, that dominate today's energy systems. These will include electricity generation from renewable sources of energy, such as wind and photovoltaic facilities, storage technologies, such as various battery systems, hydrogen storage and compressed air reservoirs. The electrical transmission network must be adapted to the new and decentralized structure of the energy systems. In addition, technologies for the generation of electricity from biomass, such as biogas or bioethanol, and from renewable electricity (power-to-gas) also need to be considered. The energy systems will also change with respect to the consumer: just a few examples include electrical vehicles, LED lamps, and new technologies in industry such as super-conductive magnetic heating systems for the processing of base metals.

Numerous different technologies must be evaluated to estimate the raw material requirements for the transition to the new energy systems, and assumptions must be made about the scope of the future utilization of each technology. Possible designs for the electricity supply in 2050 have been investigated by a working group in the "Energy Systems for the Future" project [1]. The study included both renewable and non-renewable electrical generation technologies, various types of storage technologies, expansion of the transmission net, and flexible loading (demand-side management). The study concluded that there are numerous different possibilities to conceptually plan the supply of electricity for similar total costs. Political and social issues, such as the rejection of carbon capture and storage (CCS),[1] attempts to reduce the dependence on energy imports, or a preference for small, decentral plants, may also impact the planning. It is therefore by no means clear which technology will prevail in the new systems. The study only considers 2050 as the target year for the

---

[1]This refers to the technical separation and storage of CO_2, for example from the emissions derived from power stations based on fossil fuels such as coal or natural gas. The process was investigated at a scientific pilot project in Ketzin, near Berlin.

© Springer International Publishing AG, part of Springer Nature 2019
F.-W. Wellmer et al., *Raw Materials for Future Energy Supply*,
https://doi.org/10.1007/978-3-319-91229-5_5

energy transition, but does not expand on the transitional path to reach that objective. It is therefore not clear when, in what form, and which new facilities must be constructed for the various studied scenarios. This information is however required to evaluate how the requirements of raw materials could develop with time. Based on this study, it is therefore not possible to reach any conclusions about the rate at which the raw material demands will change. This also means that no estimates can be made about the timing of possible shortages of raw materials required for each of the scenarios.

However, various studies have been undertaken more recently that focus on the raw materials required for the new energy systems. The most important and also the most recent studies are those from the US Department of Energy (DOE) from 2010/2011 [2], the Institute of Energy and Transport (JRC-IET) of the European Commission from 2013 [3], and the KRESSE study by the Wuppertal Institute from 2014 [4]. The results from the studies are discussed below.

A comparison of these studies emphasizes that a total of 45 technologies are regarded by one or several institutions as relevant. These are listed in Appendix A in Table A.1. The key thematic issues are in the fields of renewable generating technologies, storage technologies, electro-mobility, and the production of energy fuels. In contrast, energy management and waste disposal were not taken into consideration because these are not relevant to the raw materials requirements for energy systems since those raw materials that are potentially critical are not required in significant amounts, if at all, for these sectors.

The JRC-IET and the Wuppertal Institute reports both presume one political objective that the EU/Germanenergy sector should become independent of fossil combustion fuels (decarbonization), and the transformation of the German energy supply system. The conclusions pertaining to critical raw materials are derived from these objectives. The US Department of Energy study uses the reverse path. Based on the study by the US National Research Council on the raw materials critical for the economy of the USA [5], the DOE study then evaluates which sectors in the American energy sector are affected by the 16 critical elements that are defined by the NRC. These are the nine rare-earth elements lanthanum, cerium, praseodymium, neodymium, samarium, europium, terbium, dysprosium and yttrium as well as lithium, manganese, cobalt, nickel, gallium, indium and tellurium. These elements were then studied for their relevance as critical raw materials, for example the rare-earth elements in catalysts in crude oil refineries were analyzed in detail.

The demand for a raw material must be evaluated with respect to its availability in order to assess its criticality. The studies also treat this issue from different perspectives: the JRC-IET study [3] compares the expected supply of the raw material in the period from 2010 to 2030 with the amount of that raw material required if the EU states increasingly focus on non-fossil fuel energy generation (decarbonization path; Fig. 5.1). The KRESSE study [4], or the subsequent publications of the Wuppertal Institute [6] in which their criteria have been only slightly modified, use the current known reserves of a critical raw material as the baseline for their calculations, and not the expected supply situation in the future.

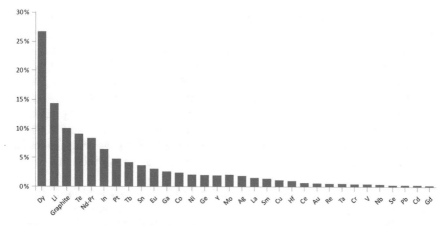

Fig. 5.1 The average expected requirement of the EU for 32 high-technology metals considered important for regenerative technologies in the period 2020–2030 as compared to their global availability (expected average supply for the period 2020–2030) [3]. Source Moss et al. [3], with permission of the EU-JRC

Both the studies from the JRC-IET and the Wuppertal Institute demonstrate that different potentially critical raw materials are required for diverse technologies. The Tables B.1–B.5 (Appendix B) cogently demonstrate the quantities of the individual raw materials that are required for the key technologies in the new energy systems such as photovoltaic facilities, fuel cells, electro-vehicles and battery systems. It must be considered, however, that not only the specific quantities of raw materials required for a specific usage are significant, but the scale of the absolute quantities of a raw material relative to its total consumption is more important for the construction of the new energy systems. Therefore, the scale of the various technologies to be installed must be known.

5.1   Demand-Driven Competition of Critical Raw Materials for Energy Technologies

As already discussed, there are practically no elements that are exclusively used for one high-technology application. Usually there is a competition between different technologies for the utilization of an element, and this also affects the new energy systems of the future. It is therefore important to recognize the level of importance of the raw material demand for the new energy systems as compared to other sectors in the national economy that also have a requirement for raw materials. The Wuppertal Institute identified critical raw materials based on a comparative analysis in order to approach the question of the criticality of a raw material for an emergent energy system of the future. Subsequently these critical raw materials were correlated with their applications in new important energy technologies and their potential development possibilities (based on selected energy scenarios) up to 2050. The demand for a raw material was defined along a possible technology development path, and finally assigned relative to today's market conditions. Germany, based on a globally just distribution, would be entitled to raw materials in a ratio relative to its proportion of the global population, which is about one percent. The Wuppertal Institute derives the so-called "one percent rule" for the proportion of the global production and the currently known global reserves of natural resources that is attributable to Germany. Furthermore, it is presumed that the demand for raw materials during the relevant period from 2011 to 2050 will be met without any problem for most of the raw material.

The opinions and criteria that support the analysis of the Wuppertal Institute, that relate to both the production as well as the reserves, can be summarized as follows: the classification of the potential critical raw materials does not only depend on the estimate of the demand for the raw materials during this time, but also on the increases in production of the mined raw material as the dynamics of its reserves. The development trends of these values over the relevant 40-year period from 2011 to 2050 are associated with significant uncertainties. So that one uncertain figure is not compared with two other uncertain figures, everything is relativized by a filter method with the known current reserves and the current production:

Step 1: it is established if the cumulative requirement over 40 years for one raw material is greater than 40% of the current global annual production. This percentage value is derived from the product of the German proportion of one percent and the time under consideration of 40 years.

Step 2: It is assumed that ten percent of the global reserves are available for renewable energies. If this is transferred to the one-percent rule, it means that 0.1% of the global reserves would be available to Germany over the next 40 years for renewable technologies. A value of 50% for renewable energies, or 0.5% for Germany, were also included in the study for a sensitivity analysis.

The criteria are usually analyzed consecutively by the Wuppertal Institute. If one of the elements does not pass the first step, then it is subject to a detailed analysis in the second step. The results of these investigations based on the second step are summarized in Table 5.1.

The reason that some of the elements, despite an overall positive forecast for future supply of rare materials, can become critical is best illustrated by the example of nickel. Nickel is required today for, among other uses, the storage of electricity in accumulators and batteries, and especially as an important alloy element in steel. Up to 2050, however, nickel will become increasingly important for the nickel electrodes in electrolysis plants which should in future, with electrical input, be used to produce hydrogen that can be stored as an energy fuel. The demand for energy storage depends on various factors, such as for example the type and degree of fluctuations in the generation of electricity. According to the assumption in the scenario with the maximum use of storage, a cumulative nickel requirement of 176,000 tonnes is expected for Germany up to 2050. This corresponds to nine percent of the annual production from 2011 and 0.23% of the global reserves. Although the first criterion

Table 5.1 Assessment of criticality of specific raw materials, subdivided according to their technical applications in renewable energy [4, p. 205ff]

Renewable energy sector	Raw material	Proportion of currently known reserves, depending on scenario (%)	Criticality based on the "1-percent" rule	Critical/Non-critical?
Wind power	Neodymium	0.005–0.094		Non-critical
	Dysprosium	0.02–0.5		Non-critical[a]
Photovoltaic	Indium	0.2–1.2 to 1.7–8.6	Critical	Critical
	Gallium	0.00014–0.0009 to 0.0016–0.009		Non-critical
	Selenium	0.013–0.07 to 0.12–0.66	Unsure	Critical?
Electricity storage	Lithium	0.024–0.48		Non-critical as compared to demand in other sectors
	Vanadium	0.58–1.16		Critical
	Nickel	0.23		Critical
	Potassium	0.00047		Non-critical
	Lanthanum and Yttrium	0.07–0.08		Non-critical

The values in the third column relate to the requirement for the reconstruction of the energy system in Germany by 2050

[a]The figures from by the Wuppertal Institute are derived from the scenarios and are presented as a range. This is marked with "?" since this sector is considered to be non-critical by the Wuppertal Institute, although the upper value of 0.5% is above the level of the critical threshold of 0.1%

is fulfilled (proportion of the global production less than 40%), the reserve criterion was checked. Relative to the global reserves the 0.1% threshold that is described above is exceeded, and therefore nickel is classified as critical [4, p. 224f], additionally: (p. 98, 200f). It should also be taken into consideration that nickel has other energy-relevant applications such as in accumulators. However, these are not considered in this storage scenario (see Appendix B, Tables B.3 and B.5), although this would further increase the critical relevance of nickel.

The thresholds defined by the Wuppertal Institute appear to be very low, but are correct if it is assumed that the cover should not cause any disruption of the market equilibrium that could initiate the feedback control cycle of raw material supply. Even a small additional demand—often only a few percent—can result in major swings in the prices on the commodity markets.[2] The Wuppertal Institute study is therefore based on a conservative approach. If short-term price swings are acceptable, then it is not necessarily problematic if disproportionately large quantities of raw materials are required for the new technologies. This would result in a change and new regulation of the commodity markets, and the reserves would increase according to the dynamic of the system.

---

[2]This is illustrated by the example of sulfur in Kesler [7, p. 106].

5.2 Response Capacity of the Global Raw Material System

With respect to the requirements for raw materials for the new emerging energy systems of the future, it is important to recognize which raw materials could become critical. It must also be considered how quickly the global raw material system can react to new demands for specific raw material that result from the reconstruction of the energy system. The expression "global raw material system" in this context refers to both the supply and the demand side in terms of the feedback control cycle of raw material supply (see Sect. 2.4). Price peaks in the past were only of short duration, and this indicates that the world raw material system could react relatively quickly to price spikes. The following two examples emphasize this:

1. China is the dominant supplier of rare-earth elements with more than 95% of the global mine production. In the past, the country exploited this strong position several times. In 2009 the export quotas were reduced by 12% and in 2010 by 40%. The prices for rare-earth elements increased sharply, and in the extreme case of dysprosium by a factor of 100 [8]. One reason for this was price speculation that always occurs during such a demand situation. Since China is by far the largest supplier, the reduced export quotas resulted in an unsecured requirement outside China of 15–20% [9]. This situation lasted for two years before it had more or less normalized (Fig. 5.2). This normalization is because the demand was in effect not so high as the speculators had anticipated. Furthermore, the price normalization was also supported by the possibilities for substitution of rare-earth elements in specific technologies. Finally, new mine production capacity was developed outside China that, despite the relatively small size, contributed to meeting the demand. In the meantime, one of these new mines has already closed.
2. During the past decades, the industrialization of China has been very dynamic. As a result, the domestic demand for raw materials has increased enormously so that China has now become the biggest consumer in the world for all the significant raw materials except for oil and natural gas. A global boom in commodity prices occurred as early as 2003 because of the increasing demand from China. The mining industry reacted to this situation by expanding their production capacities, primarily for iron and aluminum, which are the metals that are used in the greatest quantities in the world. The production of iron ore increased by a factor of 3.2 from 2001 to 2011, that of bauxite, which is the aluminum ore, increased by a factor of 1.8. The possible critical raw materials that are discussed in the following sections are, in terms of the quantities used, niche products as compared to iron ore and bauxite.

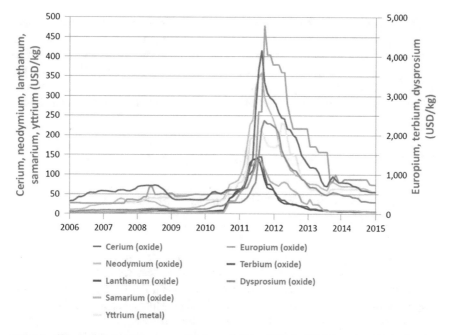

Fig. 5.2   Price trends for the rare-earth elements from 2006 to 2014 (Based on BGR [8]). The curves emphasize the "price explosion" that was caused by the reduction of the Chinese export quotes and the increasing demand for raw materials by China after 2009. The prices sank again as the excitement calmed down

Demand-driven competition of critical raw materials for energy technologies

The criticality of a raw material, with respect to the reliability of its supply, is derived from the ratio of the demand and supply of the raw material. This ratio can always change so that studies on the reliability of supply of raw materials can only provide a view of the current situation.

The designation that a raw material is critical, or could become critical, can lead to precautionary measures so that disturbances, which can lead to price peaks and initiate the mechanisms of the feedback control cycle of raw material supply, do not affect the market equilibrium, and therefore a predicted supply risk should not occur. Good analyses and additional precautionary measures to ensure a reliable supply are required to avoid the irregularities and disequilibrium in the markets, and related incidence of price peaks.

The global raw materials system is very flexible and can react relatively quickly to sudden changes in demand, presuming it is possible to expand the existing capacity and thereby avoid the long lead-times for new mining projects. There are therefore generally no concerns about prolonged shortages to the supply of raw materials. The causes for short—to medium-term shortages on the supply-side could be insufficient qualified personnel (engineers) and equipment, lengthy permitting requirements as well as delayed or inadequate investments during times of excess supply, according to the principle: "today's surplus is always tomorrow's deficit". Price swings are an important factor in initiating the mechanism of the feedback control cycle of raw material supply. There are also relevant adjustment mechanisms on the demand-side.

5.3   Critical Raw Materials for the Energy Transition

5.3.1   Raw Material Criticality Studies: Comparison of Meta-studies

During the recent years three meta-studies, which comparatively analyze numerous criticality studies, have been published. Individually these are the investigation by Erdmann and Graedel of seven criticality studies [10], the KRESSE study by the Wuppertal Institute [4], and the report of UKERC [11] that compares eleven studies.

The criticality is ranked in all three meta-studies (Table 5.2). In addition, all three studies refer to the report of the National Research Council [5] about critical raw materials for the US economy as well as the EU report of [12] on the EU-14 critical raw materials. The studies selected by Erdmann and Graedel investigate the general economic importance of raw materials in the EU, Germany, Great Britain, Japan, Korea and the USA as well as Bavaria, which is a Federal State in Germany, during the period 2008–2010. The KRESSE report of the Wuppertal Institute reviews studies from the period 2007 to 2012 for Germany, the USA and the EU that are relevant for the transformation of the energy supply system in Germany. The UKERC meta-analysis reviews studies from the period 2007 to 2011 that are the US-American NRC and DOE studies [2, 5], EU studies and other studies that were especially relevant to the United Kingdom (UK). The UKERC uses a standardization process that in effect limits the number of studies that can be considered. On the basis of this standardized comparison, nine raw materials were considered to be particularly critical for the renewable energy technologies of the future.

Among others, the studies by the Institute for Futures Studies and Technology Assessment (IZT) and adelphi [13], which classifies germanium, rhenium and antimony as critical raw materials for Germany, were not included in the three meta-studies. These studies did not only review the raw materials that are relevant to the energy systems, but also those that are important for the whole of the German economy. Studies [14–17] that recommend actions for research and development without developing their own list of critical raw materials, which is a necessary condition for a comparison of different criticality evaluations, were also not included in the three meta-analyses. The Resnick study, for example, is based on the critical raw materials that were defined by the US Department of Energy [2].

Table 5.2 Comparison of the meta-studies by Erdmann and Graede [10], Wuppertal Institute [4] and UKERC [11]

Study	Erdmann/Graedel [10]		Wuppertal Institute [4]		UKERC [11]	
Reference framework	Overall economy		Energy systems		Energy systems	
Criticality Ranking	0–4	Ratio	1–7	Ratio	0–10	Ratio
Niobium	4	5/7	(2)	1-2/12	(Not considered)	
All platinum metals (not subdivided into individual metals)	(See below 4–3)		(2–3)		**6, 7**	2/11
Ruthenium	4	2/2	(2)	1–2/12	(Not considered)	
Rhodium[e]	4	2/2	(2)	1–2/12	(Not considered)	
Platinum[e]	4	6/7	(3)	3–4/12	(Not considered)	
Tungsten[c,d]	4	5/6	(2)	1–2/12	(Not considered)	
Indium[b,d]	4	5/7	**6**	9–10/12	**6**	8/11
All rare-earth elements (not subdivided)[a,c,d,e]	4		**(Considered separately)**		**7**	5/11
Scandium[b]	4	3/3	(3)	3–4/12	(Not considered)	
Yttrium[a,d]	4	5/5	(4)	5–6/12	(Not considered)	
Neodymium[c]	(See above REE)	6/6	**7**	>10/12	(Not considered)	
Dysprosium[c]	(See above REE)	4/4	**6**	9–10/12	(Not considered)	
Praseodymium[c]	(See above REE)	3/3	**5**	7–8/12	(Not considered)	
Terbiuim[d]	(See above REE)	5/6	**5**	7–8/12	(Not considered)	
Tellurium[b,d]	(2)	0/5	**5**	7–8/12	**5**	5/11
Silver[a,b,c,d]	(2)	1/4	(4)	5–6/12	**4**	5/11
Cobalt[a,c,e]	(3)	3/6	(4)	5–6/12	**2, 7**	5/11
Gallium[b,d]	(3)	3/6	(4)	5–6/12	**7**	5/11
Germanium[b]	(3)	3/6	(4)	5–6/12	**5**	4/11
Lithium[a,d]	(3)	3/6	(3)	3–4/12	**4**	3/11

The results are shown in detail in Figs. B.1–B.3. **The raw material that is ranked as most criticalin each of the studies is shown in bold letters**. In order to compare the studies, the ranking of the raw materials in the other studies is shown in brackets. The reference framework for each of the meta-analyses is provided in the second row. The range of each of the criticality rankings is provided in the third row of the table. The ratio refers to the number of raw materials defined as "critical" relative to the total number of the individual studies that were taken into consideration, and for Erdmann and Graedel only for the number of studies in which the raw material was actually investigated. The ratio is subdivided differently in the meta-analyses (the Wuppertal Institute only provides ranges), and dependent on the focus of the individual studies. Only the Erdmann and Graedel study evaluated if, for example, the platinum group metals or rare-earth elements were considered as a single raw material or as a constituent of a group of raw materials. The table combines the highest category (greatest level of danger for the supply of the raw material) from the Erdmann and Graedel analysis (comparison of seven studies) and the three highest categories in the KRESSE analysis (comparison of twelve studies) by the Wuppertal Institute. The raw materials were preselected in the UKERC analysis (comparison of eleven studies), and are provided in the table. All the studies in each of the meta-analyses are listed in Appendix 3
Examples of use in the energy sector [4]. [a]Batteries, [b]Photovoltaic, [c]Motors/generators/turbines, [d]Applications for electricity, [e]Catalysts/refining

The KRESSE study by the Wuppertal Institute is the only one of the three meta-analyses to include a report from the University of Augsburg [18] that provides a list of "sustainability factors" for every raw material. The reserves, trade, environmental impacts, processing and smelting, possibilities for substitution and possibilities for recycling were all investigated. There is no overall ranking of the raw materials, and therefore it cannot be compared with the other studies on criticality.

Evaluation of criticality

The rare-earth elements and the platinum group elements that are most often used in applications are critical in the seven studies reviewed by Erdmann and Graedel [10] (Table 5.2; Appendix B, Fig. B.1). The electronic metal indium, the steel alloy metal niobium, and the refractory metal tungsten were also classified as critical. Each of the rare-earth elements were investigated separately in the twelve studies (Table 5.2; Appendix B, Fig. B.2) reviewed by the KRESSE report [4]. Neodymium, because of its use in magnets in wind power facilities, was classified as by far the most critical element, followed by dysprosium, terbium and praseodymium. Since the KRESSE report only investigated renewable energies, the platinum group elements, which are used in catalysts and fuel cells, are not included. Indium and tellurium are also classified as very critical.

The standardized comparison in the UKERC meta-analysis shows how many of the eleven criticality studies include each of the raw materials considered to be critical (Appendix B, Fig. B.3). In this case, it must be emphasized that a raw material is not automatically more critical just because it is listed in more studies as being critical. For example, the by-product raw material gallium is only categorized as critical in five of the criticality studies investigated by UKERC, but is still the most critical raw material because gallium is required in a number of advanced technology applications that extend beyond the energy technologies [19–21]. Gallium is followed on the normalized criticality scale by the rare-earth elements, the platinum group elements, and then indium, germanium and tellurium. Five of these metals are by-product elements, whereby silver that is listed as a single raw material, is often recovered as a by-product of lead and zinc or copper production and can therefore considered to be a by-product, too. Indium is listed most times as critical in most of the studies (8 from 11), but only has the fourth highest criticality value.

There are numerous specific issues in each of the investigated studies. For example, phosphorus is listed as critical in the study of Bavaria [22], which was integrated into the Erdmann and Graedelmeta-analysis, and this is most probably a reflection of the importance of agriculture in Bavaria. This is the only study from this period that classifies phosphorus as being critical. Phosphorus is also currently included in the EU-20 critical raw materials [23], but is not considered in the meta-analyses. The reason for listing phosphorus as being critical is not based on a supply issue that can be identified today, but primarily because phosphorus is essential for plant growth and cannot be substituted.

The DERA Studies

The Mineral Resource studies from the German Mineral Resources Agency (DERA) have contributed significantly to the discussion in Germany about criticality. The first study [24] from 2012 evaluated the criticality of 35 primary mineral resources (metals and industrial minerals). The global supply concentration was investigated by means for the Herfindahl-Hirschmann index with respect to the concentration of production in countries, the country risks were evaluated by World Bank indicators, and the global concentration of company activity was also partly taken into consideration. The conclusions of this study on the criticality of supply of raw materials were similar to the EU-14 list (Compare EC [12]) of the EU Commission. In their second study from 2015 [25], the DERA investigated as many as 61 primary raw materials and expanded its analysis to include more than 213 raw material products further up the value chain. The latest DERA study demonstrates that those raw materials for which there is a strong concentration of supply (in mining and exporting countries with moderate to high country risks), are particularly critical both at the mining stage as well as for the numerous intermediary products in the value-added chain.

5.3.2 Comparison of Analyses of Critical Raw Materials Required for the Energy Transition

After the evaluation of the meta-analyses that integrate different studies, three of the most important resource studies in the world (the study from the US Department of Energy (DOE) [26], the study by the Joint Research Center at the Institute of Energy and Transport (JRC-IET) in the European Commission [3], and the KRESSE study from the Wuppertal Institute [4] are reviewed with respect to the results specifically pertaining to energy systems of the future. Whereas the KRESSE study was completed on behalf of the German Federal Parliament, the JRC-IET and DOE studies are also emphasized since they are included in the three important meta-analyses that are discussed in this chapter. The DOE study is the oldest one of this type, and includes all the energy technologies, and not only the renewable energies. However, all the raw materials that are classified as critical in these studies are also relevant to the energy systems of the future (Table 5.3).

The JRC-IET study classified a list of 32 key metals according to a combination of market and geopolitical factors. The market factors include, for example, constraints on the expansion of supply or the probability of a strong increase in demand. Geopolitical factors include the concentration of the supply in specific countries or the political risk in the producer countries (Table 5.4).

Finally, the KRESSE study [4, p. 241f], which basically evaluated the geological availability, comes to the following conclusions:

1. "The geological availability of mineral raw materials does not present a constraining factor on the planned expansion of renewable energies. However, it is possible that not all the technological types can be universally implemented".
2. "With respect to the supply of the mineral raw materials, some components or sub-technologies of wind power, photovoltaic as well as battery storage were identified as being potentially "critical" (….). However, non-critical alternatives are available for these technologies that could be increasingly used or are already today dominant in the market".

Table 5.3 Time scale for the evaluation of the criticality of raw materials according to DOE [26]

Short and medium-term ranking of criticality (in years)					
Critical		Nearly critical		Key material, but not critical	
(0–5)	(5–15)	(0–5)	(5–15)	(0–5)	(5–15)
Yttrium	Yttrium	Cerium	Lithium	Lithium	Indium
Neodymium	Neodymium	Lanthanum	Tellurium	Nickel	Nickel
Dysprosium	Dysprosium	Tellurium		Cobalt	Cobalt
Europium	Europium	Indium		Gallium	Gallium
Terbium	Terbium			Manganese	Manganese
				Praseodymium	Praseodymium
				Samarium	Samarium
					Cerium

The DOE takes different time lines into account for the evaluation of the supply and criticality of raw materials for energy systems of the future

Table 5.4 Ranking of criticality for 32 of the raw materials that are important for energy systems by the Joint Research Center Institute for Energy and Transport [3]

High	Moderate to high	Moderate	Moderate to low	Low
REE: Dysprosium, Europium, Terbium, Yttrium	Graphite	REE: Lanthanum, Cerium, Samarium, Gadolinium	Lithium	Nickel
REE: Praseodymium, Neodymium	Rhenium	Cobalt	Molybdenum	Lead
Gallium	Hafnium	Tantalum	Selenium	Gold
Tellurium	Germanium	Niobium	Silver	Cadmium
	Platinum	Vanadium		Copper
	Indium	Tin		
		Chromium		

Critical Raw Materials for the Energy Transition

Many of the rare-earth elements as well as the platinum group metals, indium, niobium, tungsten, gallium, germanium and tellurium are raw materials that are most commonly classified as critical in their supply.

Criticality studies can reach different outcomes depending on the use and assessment of market indicators as well as the objectives. Nonetheless these studies are an important contribution to the discussion about possible bottlenecks in the supply chain. The most important aspect of these studies is to identify price and supply risks on the commodity markets as early as possible, and to continue regular detailed analyses. The decision-makers in politics, business and the society can therefore be provided with important information, on the basis of which they can react to future changes in the supply chain and take the necessary measures to ensure the supply of these raw materials.

5.4   Author's Assessment of Criticality

5.4.1   Raw Materials Derived from Individual Deposits, in Particular the Rare-Earth Elements, and By-Product Elements

An evaluation of the DOE study [2], the JRC-IET study [3] and the meta-analyses by Erdmann and Graedel [10], the Wuppertal Institute [4], and UKERC [11] demonstrates that, together, they come to very similar conclusions. The rare-earth elements are classified as critical in all these studies, and tellurium and indium are classified as critical to almost critical.[3] The large number of by-product metals is noteworthy, and the critical and almost critical raw materials are subdivided further according to how they occur in mineral deposits—those that occur in their own individual deposits and those that are by-product elements (Table 5.5).

The supply of raw materials from their own individual ore deposits regulates itself in the market economy systems according to the feedback control cycle of raw material supply (Compare Bast et al. [27]). The analysis of raw materials, such as those made by the German Mineral Resources Agency (DERA) for the Federal German Parliament,[4] provide important systematic basic information and practical examples. Monitoring of raw materials, for example the DERA Raw Material Lists, involves

---

[3]The JRC-IET study (Moss et al. [3]) subdivides the criticality into "medium", "medium to high" and "high" classes. The "medium to high" class is the same as "almost critical".

[4]CDU/CSU/SPD [28]: With the title "monitoring the expansion", the following is written „we will commission the German Mineral Resources Agency to monitor the critical raw materials and regularly report on the availability of those raw materials that are critical for the German economy".

Table 5.5 Occurrence of raw materials that are classified as critical and almost critical into by-products or their own individual deposits [2, 3]

Type of occurrence of the raw materials	Relevant study		
	DOE	JRC-IET	DOE + JRC-IET
Occurs in own individual mineral deposit	Cerium[a], Lanthanum[a], Lithium	Praseodymium[a], Graphite, Platinum	Yttrium[a], Neodymium[a], Dysprosium[a], Europium[a] Terbium[a]
Occurs as a by-product		Gallium, Rhenium, Hafnium, Germanium	Indium, tellurium

[a]Rare-earth elements

observing both the high-level developments on the commodity markets at regular intervals and, increasingly, the supply, demand and prices at the initial stages of processing of the raw materials (compare Sect. 2.3). Based on this initial overview, the raw materials or groups of raw materials, and their intermediary products that are classified as being potentially critical are evaluated in more detail in special risk assessment reports. For example, copper, tin, antimony, tungsten, bismuth, the platinum group metals, and lithium have already been analyzed separately, and this makes it easier to evaluate the specific market sectors. Possible weaknesses in the supply chain for these raw materials and alternative sources of supply can be identified. Furthermore, the DERA is also engaged in improving the methodology of identifying and evaluating the price and supply risks, and thereby improving the knowledge for and expanding the supply of these raw materials and intermediary products to Germany. DERA distributes the results, and their methodology for assessing criticality, as publications and information events that are freely available to the public, although the focus is towards the relevant businesses as they provide important information about supply shortages that could possibly occur as well as about new supply sources [14, 24, 29–31].

Prevention by political intervention?

In the case of critical supply situations, the events that caused this circumstance to arise must be analyzed in detail. For example, they could be related to a peak in demand or political interests. Political measures could either provide the necessary preventive support, or be more direct. The actions taken by Austria in the case of the Mittersill mine are mentioned here as an example of possible actions. Despite a slump in the tungsten prices, the Mittersill tungsten mine in Austria could be maintained on a care-and-maintenance basis with state assistance during 1992/1993, and is today one of the more important tungsten producers in the EU.

This poses the question: is it justified to maintain a mine in a market economy with state financing? The basic difference between a normal industrial organization (such as an automobile factory) and a mine must be clarified: basically the same input factors such as productivity, materials, equipment or human knowledge can be created or acquired for any normal industrial business, wherever it might be located. This situation is quite different for a mining operation, which must be located at the mineral deposit. Furthermore, every mineral deposit is "unique": defined by different values for reserves, grades and numerous other parameters that control the economics of extraction. These parameters can be summarized as the "credit rating" [32]. The "credit rating" of a mineral deposit is based mostly on the geology, and as a consequence there is a merit ranking (from low-cost to high-cost mines) for each mineral raw material that can only be improved by better management, rationalization, etc., to a certain degree. Because, in contrast to industrial products, the raw material prices vary like a temperature curve (compare Sect. 3.2), opportunities then arise for companies and countries with low-cost mineral deposits, sometimes supported by calculated price-dumping, to use the periods of low prices to establish monopolies or oligopolies. The question therefore arises: should a raw material consumer country such as Germany not take action, particularly in the case of critical raw materials or raw materials of strategic economic importance that are vital for high-technology products and for the energy systems of the future, to protect itself from the formation of such monopolies or oligopolies that are based on geological, and not economic, advantages.

For example, there are conceivable situations in the future whereby competitive mines can be forced to close because of price-dumping. Under these circumstances, it can only be assumed that the customers of raw materials will purchase their requirements to a limited extent from an alternative but more expensive producer as a way of diversifying their possible sources, as is a sort of "insurance premium". These customers must also compete with their products on the world markets, and are therefore only able to offset to a certain extent the higher prices for raw materials. It is conceivable that concerted actions, for example in the form of state subsidies, also at the international level—such as the EU, are required to ensure that producers are not forced to close their mines as a result of market irregularities, but can maintain their mines on care-and-maintenance. Production from these mines can be reactivated as soon as the prices rise again, and the monopoly situation can thus be avoided. In turn, the customers would be required to make long-term supply contracts prior to the support measures, which would provide the end-user with protection from possible monopolistic situations, and the mines with support towards keeping them open.

Complex adjustment of production rates: the example of the platinum group metals

The feedback control cycle of raw material supply only applies to a limited extent to the by-products (Sect. 2.4, Box I). There is a significant lack of information about the future supply and market transparency of the by-product raw materials. For example, there are repeated warnings of a shortage of indium because the reserve situation is very uncertain. The standard publication from the US Geological Survey on mineral reserves, the *Mineral Commodity Summaries*, states "quantitative estimates of reserves are not available" [33].

The platinum group metals[5] provide an example that emphasizes the complexity of the occurrence of mineral raw material and the adjustment of rates of production to compensate for changing requirements. Platinum group metals are important to produce catalysts and in electrolysis for extraction of hydrogen. They could be of importance in the storage technologies (power-to-gas) for the energy systems of the future. Iridium should be emphasized in this respect because it cannot be substituted for some of its uses, even by other platinum group metals.

The feedback control cycle of raw material supply will function if there is a sudden increase in the demand for iridium. The iridium price, which has up to now usually been lower than the platinum price, would increase and result in a motivation to increase the production of iridium. This additional demand can be relatively quickly covered by the so-called intermediates, which are intermediary products formed during the separation processes during the production of platinum. The six platinum metals all occur together and, because platinum has always been the most sought after of these two metals, iridium is a by-product of the platinum production. The intermediates are either stockpiled or processed further by the platinum smelter operators, depending on the demand for each of the individual elements. Since the separation of iridium, ruthenium and osmium occurs at the end of the process, they are initially stockpiled as a constituent of these intermediates.

An increase in the iridium demand can also be covered by exploiting mineral deposits, or sections of deposits, with relatively high iridium content. Generally, the deposits are exploited because of their high platinum content that commands particularly high prices. However, if the iridium price increases very strongly because of the demand, it could be possible to switch to those deposits with higher iridium grades and thereby produce more iridium, although the quantity of produced platinum would decrease. The Merensky Reef in the Bushveld Complex, South Africa, contains the largest known deposits and is the main source of the platinum group metals; the iridium-platinum ratio is 1:50 in the Merensky Reef. The UG-2 Reef is another deposit in the Bushveld Complex, and the ratio is 1:20 and therefore iridium is two-and-a-half times more abundant. If there is an increase in the iridium price, the UG-2 Reef will be more intensively exploited. The case whereby one of the by-product metals becomes a dominant driver can be illustrated by the production of the rare-earth elements. Initially europium was the main driver, and today it is mostly dysprosium—not in quantity, but because of the increased demand and therefore higher price.

If the iridium price increases strongly, then it could become attractive to explore for those types of mineral deposits with higher iridium contents and bring them into production. These would be mineral deposits that would today be regarded as "un-

---

[5]Similar to the rare-earth elements, the six platinum group metals (see Sect. 1.3) always occur together in mineral deposits. The ratios of these metals vary according to the type of mineral deposit. The most important of these metals are, by far, platinum with a global mine production of 190 tonnes per year, and palladium with 210 tonnes per year. The other platinum group elements are only produced at a low level, for example only three to four tonnes of iridium per year. The production is mainly from their own individual deposits, and to a lesser extent as by-products from nickel deposits (see Fig. 3.14).

conventional" from an economic viewpoint. Mineral occurrences with a much higher relative content of iridium due to geochemical enrichment processes are known, for example, from ophiolites [34, p. 587].[6]

The potential of the technosphere must also be taken into consideration together with the primary production of raw materials. With respect to the platinum group metals, specifically iridium, it is important to note the technically good recycling capability of PGE-bearing catalysts used in hydrogen electrolysis. Similar to the PGE catalysts in the petrochemical industry, the industrial cycles can be designed so that only very minor losses occur during the life cycle. Initially the application of new technologies, such as electrolysis facilities in which iridium is increasingly used as a catalyst, results in an increasing demand for iridium that must be covered by production from primary mineral deposits. This ultimately results in the establishment of a stock of iridium reserve in products in the technosphere. The platinum group metals can be (and are) continually recycled from these products at the end of their life cycle.

Creating transparency in niche markets

The boundaries between reserves, resources and geopotential are dynamic. With respect to the focus on the evaluation of future supply situation of raw materials, the objective of raw material analyses must be to identify the relevant geopotential that will become available in the future. However, the geopotential is the "big unknown", and cannot yet be estimated with today's technical requirements. Previous experience shows that the market demand could always be covered, although the group of raw materials that occur as by-products are a special case. Even the reserves and resources of by-products are difficult to define. The example of germanium has recently provided a more precise estimate [35, 36].

The market for the by-product raw materials, and also for the rare-earth elements, is very restricted, and is often termed a niche-market. The trading is not normally performed on the major exchanges, but is restricted to a few commodity producers and a small number of customers. These markets are often characterized by a significant lack of market transparency. The threshold[7] to enter the market is correspondingly very high [37]. The entry threshold for possible new producers could be reduced by increasing the transparency in the market. The direct communication between the producing and processing companies is very important to create transparency in the market. The exchange of information could be managed by an international study

---

[6]Ophiolites are rock complexes formed from the oceanic crustal rocks or oceanic lithosphere, and are now found on land due to plate tectonic processes.

[7]The threshold to enter the market is reflected by the degree of difficulty for a mining company to find a market for its products, which is an essential requirement for financing the project. For example, there is no threshold for a gold mining company to enter the market because gold is cash (gold mining = cash mining), and the product can be immediately exchanged for cash by the banks. The metals that are traded on the exchanges, such as copper or zinc, are always marketable and therefore have a very low entry threshold. The markets for the rare-earth elements or by-product raw materials, such as indium or germanium, are niche-markets where a few customers control the trading. The threshold for mining companies to enter these markets is very high.

group of consumers and customers. This type of so-called Commodity Study Group is already functional for lead, zinc, copper and nickel.

In this context, it is very important that, to ensure that Germany has a reliable supply of raw materials in the future, the political and industrial protagonists are prepared to provide sufficient research opportunities and funding. Research and development are the basis for the development of market-ready mitigation strategies, alternative production methods, efficient production designs as well as for improving the possibilities for recycling. The issue of social acceptance will always remain to be an integrative component that is increasingly important for successful and practical implementation, not least because in the case of the energy systems of the future the classical consumer has often become a "prosumer".[8]

5.4.2   Phosphorus and the Noble Gas Helium

The noble gas helium and element phosphorus should be addressed in more detail in any discussion on the long-term security of raw material supply, because the usual self-regulation of the market economy does not apply to either helium or phosphorus. It can therefore be expected that these raw materials will become increasingly critical for the national economy without an intervention with a long-term focused action to ensure the reliability of the supply.

Helium—lost to space

Helium for example, is not mentioned as a critical raw material in any of the criticality studies that have been reviewed (helium is first mentioned in the EU-2017 list of critical raw materials). However, in the long-term it should be considered as a potential critical material. The noble gas helium is required in large quantities for refrigeration systems in the energy technologies, for example for cooling technical facilities (cryogenic systems), and is therefore also important for the energy systems of the future [38]. A regular evaluation of this raw material is therefore required. Helium was already classified by the USA as a strategic raw material in 1925. It occurs in small amounts in natural gas deposits. Because the helium content in the natural gas deposits in southwest USA is unusually high, and because it was of critical importance for the Zeppelin airships, it has been separated from the natural gas and collected in a geological reservoir since 1925 as part of a federal program.[9] In 1996, the American Congress agreed to privatize the helium reserves. The American Physical Society (APS) and the Materials Research Society (MRS), an international

---

[8]The term "prosumer" refers to the double-function of an organization or individual that simultaneously performs the roles of a consumer and producer. With respect to the energy transition these are, for example, the individual households that own their own energy production facilities and are therefore both energy consumers and producers.

[9]The German zeppelin had to be filled with hydrogen gas because the USA did not sell the strategic raw material helium to Germany. In this content, the explosion of the Hindenburg Zeppelin on May 6, 1937, in Lakehurst, New York, is well known.

society of materials research, recommended in 2011 that the helium storage should be returned to state control [15]. The main reason for this is that, with respect to the long-term security of the raw material, the economic interests of a private operator, namely the maximizing of the profitability, would often conflict with the national economic interests to maintain a cost-efficient control over the supply of the raw material.

Helium is today produced commercially in the USA, Algeria, Qatar, Russia and Poland. Because helium is lighter than air, it is naturally a critical material that can be lost from Earth. If it is not separated during the production of natural gas it dissipates into the atmosphere, and mixes with the air where it is diluted and ultimately escapes from the atmosphere into space.

Phosphorus cannot be substituted

Phosphorus is an important nutritional fertilizer for plants that cannot be substituted by any other substance, and is therefore classified as a bio-essential element. The criticality of phosphorus relates to the higher-grade, stratiform deposits, which do not contain an unlimited resources. In contrast to the bio-essential nitrogen, which is available in large quantities as nitrogen in the air and can be recovered in more or less unlimited quantities by chemical processing (Haber-Bosch process), the terrestrial potential of phosphorus is finite. Although the technology has already been developed, an extensive recovery of phosphate in a sort of utilization cycle is currently not economically feasible. A common way to re-use this nutritional fertilizer is, for example, the application of liquid manure as a fertilizer (see Sect. 4.3.6). Politicians in the EU repeatedly demand further clarification on the future availability of phosphorus (phosphorus reserves) [39]. This situation is the only reason for phosphorus to be included in the list of EU-20 critical raw materials [23], although the EU report does not predict any supply problems in the short term—and even a large surplus beyond 2020—but the supply risk is high due to concentrated production from mainly three countries. The new analyses generally review the overall geological availability, or the geopotential. With respect to phosphorus, there is an historical precedent that could be adapted: International Geological Correlation Program Project 156 "Phosphate deposits of the world" from 1977 to 1984 [50]. This project compiled an excellent database with respect to the level knowledge at that time. A proposal has recently been made to continuously monitor the geopotential of phosphorus or phosphate at the international level [41].

5.4.3 The Metals Lithium and Copper

Lithium and copper are often the focus of public discussion about the criticality of their supply situation for energy systems. Copper is an electrical conductor, and the electricity metal per se, and lithium could be important as the metal in batteries, particularly in the electrification of the automobile traffic.

Lithium

The 2011 DOE study [26] classifies lithium as "almost critical" for the next 5–15 years (Sect. 5.3.2). Lithium is mentioned three times as a critical raw material in the comparative analysis by Erdmann and Graedel [10], it is mentioned three to four times in the KRESSE study [4], and in four of the eleven studies in the UK Energy Research Center [11] meta-analysis it is classified as critical (Figs. B.1–B.3). Furthermore, it is also classified as critical in a recent study [42] commissioned by the World Wild Life Fund for Nature (WWF). However, it is not present in the list of EU-20 critical raw materials.

The problem with analyses such as, for example, the WWF study is that the reserves and resources are treated as a static figure, rather than a dynamically changing value. The resources are even regarded as "everything that is expected to exist" [42]. As a result, the geopotential and the results of any future exploration are completely ignored (see Sect. 2.2 and Fig. 2.3). There is nothing to suggest that the feedback control cycle of raw material supply should not function for lithium. The study by the Fraunhofer Institute for Systems and Innovation Research (ISI) concludes that "even presuming the worst-case scenarios there is no expectation for a shortage of lithium reserves during the next four decades" [21, p. 1], see also [43].

Copper

There is also no reason to presume that there could be a shortage of copper during the next decades (Compare [44]). Copper is not listed as a critical raw material in the meta-analysis by Erdmann and Graedel, the KRESSE study, or the UKERC study (Table 5.2; Appendix B, Figs. B.1–B.3) [4, 10, 11]. However, copper is one of the most important metals that various authors repeatedly consider to be affected by constraints to its availability in the long-term, and this is then repeatedly repudiated by well-known mineral commodity economists.[10]

The discussion finally ended on the issue of whether the geopotential of a mineral raw material can be determined. The discussion about a "Peak Minerals" can generally be regarded to be inappropriate (Sect. 3.1.1) because the question about the availability of sufficient copper from both the technosphere and the geosphere must be addressed. As the situation for aluminum demonstrates (Fig. 3.26), the proportion of supply of secondary metals from the technosphere is continuously increasing. This is also true for copper, whereby copper is a relatively noble metal and has the additional advantage of being recycled to the same quality, which is not the case for aluminum. There is no down-grading of the product during the recycling process.

---

[10]There is a well-known dispute between Thomas E. Graedel of Yale University and John Tilton, one of the leading mineral economists in the world (Gordon et al. [45]; Tilton/Lagos [46]. Graedel and his supporters presume that the total resource box for copper (see Fig. 2.3) can be estimated, but Tilton disputes this approach and uses an opportunity cost paradigm that is approximately comparable to the feedback control cycle of raw material supply. A copper peak was recently discussed in Science News Focus (Kerr [40], that has been predicted by Australian researchers to occur in about 2050. This was repudiated both by Tilton and also by Schodde [47–49], who is one of the leading experts on copper.

Furthermore, possible technology advances must be taken into account. For example, the increased usage of glass-fiber cables, which commenced in the 1990's, resulted in a significant effect on the copper consumption. Similar effects could, for example, occur related to advances with super-conductivity. All these effects are taken into consideration by the market mechanisms in the feedback control cycle of raw material supply [14].

Our Assessment of Criticality

An intervention from the governmental side to protect mining operators from market distortions, for example by price-dumping or political interference, should be considered.

The creation of market transparency by commissions of international and independent experts can contribute to avoiding market distortions and supply risks, particularly in the niche-markets.

The global occurrence of phosphorus in the geosphere is extensive and, according to our current knowledge and based on the existing static range of at least 300 years, is estimated to be sufficient over the long-term, and to cover the expected demand. However, phosphorus (phosphate) cannot be substituted as a plant nutrient, and therefore requires a long-term strategy. Until now and into the foreseeable future it is not possible to close the global anthropogenic cycle, and under these circumstances the resources of phosphate are finite. There are two fields for further research. Firstly, the actual geopotential of phosphorus, or the phosphate required by the plants, must be clarified. Secondly, the technology for recycling phosphate into the secondary economy, whereby the phosphate must be recovered by means of recycling into its useful economic chain, must be established.

Because of its application in refrigeration, helium could become an important element in the implementation of energy technologies of the future. Helium can escape from the atmosphere into space, and therefore special technical procedures are required to retain and store this raw material.

No significant problems are foreseen for the supply of either lithium or copper for the "energy systems of the future". There are no indications to suggest that the mechanisms of the feedback control cycle of raw materials supply should not be sufficient to ensure that the future demand for these raw materials is covered, even if some studies present an alternative scenario for lithium based on the current reserves and resources.

References

1. Elsner, P./Fischedick, M./Sauer, D. U. (Hrsg.): *Flexibilitätskonzepte für die Stromversorgung 2050: Technologien – Szenarien – Systemzusammenhänge* (Schriftenreihe Energiesysteme der Zukunft), München 2015.
2. US Department of Energy: *Critical Materials Strategy*, Washington DC 2010. URL: http://www.energy.gov/sites/prod/files/piprod/documents/cms_dec_17_full_web.pdf [accessed: 28.10.2014].
3. Moss, R. L./Tzimas, E./Willis, P./Arendorf, J./Tercero Espinoza, L.: *Critical Metals in the Path towards Decarbonisation of the EU Energy Sector – Assessing Rare Metals as Supply-Chain Bottlenecks in Low-Carbon Energy Technologies* (Scientific and Policy Reports), Petten: European Commission, Joint Research Centre, Institute for Energy and Transport 2013. URL: https://setis.ec.europa.eu/newsroom-items-folder/new-jrc-report-critical-metals-energy-sector [accessed: 02.05.2014].
4. Wuppertal Institut für Klima, Umwelt, Energie GmbH: *KRESSE — Kritische mineralische Rohstoffe bei der Transformation des deutschen Energieversorgungssystems*, Abschlussbericht an das Bundesministeriums für Wirtschaft und Energie, Wuppertal 2014. URL: http://wupperinst.org/de/projekte/details/wi/p/s/pd/38/ [accessed: 15.12.2014].
5. National Research Council of the National Academies: *Minerals, Critical Minerals, and the U.S. Economy*, Washington, D.C.: The National Academies Press 2008.
6. Viebahn,P./Soukup, O./Samadi, S./Teubler, J./Wiesen, K./Ritthoff, M.: "Assessing the Need for critical Minerals to shift the German Energy System towards a high Proportion of Renewables". In: *Renewable and Sustainable Energy Reviews*, 49, 2015, pp. 655–671.
7. Kesler, S.E.: *Mineral Resources, Economics and the Environment*, Macmillan: New York 1994.
8. Bundesanstalt für Geowissenschaften und Rohstoffe: *BGR-Datenbank*, Hannover: Bundesanstalt für Geowissenschaften und Rohstoffe 2014.
9. Kingsnorth, D.: *Strategies for securing sustainable Supplies of Rare Earth,* presented at the Deutschen Rohstoffagentur DERA, Berlin, 20.10.2014.
10. Erdmann, L./Graedel, T.E.: "Criticality of non-fuel minerals: a review of major approaches and analyses". In: *Environmental Science and Technology*, Bd. 45: 18, 2011, pp. 7620–7630.
11. United Kingdom Energy Research Centre (UKERC): *Materials Availability. Comparison of Material Criticality Studies – Methodologies and Results* (Working Paper III), 2013.
12. European Commission: *Critical raw materials for the EU* (Report of the Ad-hoc-Working Group on defining critical Raw Materials), Brussels 2010. URL: http://ec.europa.eu/enterprise/policies/raw-materials/files/docs/report-b_en.pdf [accessed: 01.05.2014].
13. Erdmann, L./Behrendt, S./Feil, M.: *Kritische Rohstoffe für Deutschland* (Studie des Institutes für Zukunftsstudien und Technologiebewertung (IZT) und adelphi für die KfW-Bankengruppe), Berlin 2011.
14. Dorner, U./Buchholz, P./Liedtke, M./Schmidt, M.: *Rohstoffrisikobewertung–Kupfer, Kurzbericht* (DERA Rohstoffinformationen Nr. 16), Deutsche Rohstoffagentur in der Bundesanstalt für Geowissenschaften und Rohstoffe 2013.
15. American Physical Society/Material Research Society: *Securing Materials for Emerging Technologies*, Washington DC 2011.
16. Fraunhofer Institut für System- und Innovationsforschung (Fraunhofer ISI): *Energietechnologien 2050 –Schwerpunkt für Forschung und Entwicklung* (ISI-Schriftenreihe "Innovationspotentiale"), Stuttgart: Fraunhofer-Verlag 2010.
17. Resnick Institute: *Critical Materials For Sustainable Energy Applications*, Pasadena 2011.
18. Achzet, B./Reller, A./Zepf, V./Rennie, C./Ashfield, M./Simmons, J.: *Materials critical to the Energy Industry. An introduction* (Report for the BP Energy Sustainability Challenge), Augsburg: University Augsburg 2011. URL: http://www.physik.uni-augsburg.de/lehrstuehle/rst/downloads/Materials_Handbook_Rev_2012.pdf [accessed: 28.10.2014].

19. Angerer, G./Erdmann, L./Marscheider-Weidemann, F./Lullmann, A./Scharp, M./Handke, V./Marwede, M.: *Raw Materials for emerging Technologies*, Stuttgart: Fraunhofer IRB Verlag 2009. URL: http://www.isi.fraunhofer.de/isi-en/service/presseinfos/2009/pri09-02.php [accessed: 27.10.2014].
20. Angerer, G./Erdmann, L./Marscheider-Weidemann, F./Lullmann, A./Scharp, M./Handke, V./Marwede, M.: *Rohstoffe für Zukunftstechnologien: Einfluss des branchenspezifischen Rohstoffbedarfs in rohstoffintensiven Zukunftstechnologien auf die zukünftige Rohstoffnachfrage*, Stuttgart: Fraunhofer IRB Verlag 2009. URL: http://www.isi.fraunhofer.de/isi-en/servic e/presseinfos/2009/pri09-02.php [accessed: 27.10.2014].
21. Angerer, G./Marscheider-Weisemann, F./Wendl, M./Wietschel, M.: *Lithium für Zukunftstechnologien – Nachfrage und Angebot unter besonderer Berücksichtigung der Elektromobilität*, Karlsruhe 2009. URL: http://publica.fraunhofer.de/eprints/urn:nbn:de:0011-n-1233149. pdf [accessed: 28.10.2014].
22. Pfleger, P./Lichtblau, K./Bardt, H./Reller, A.: Rohstoffsituation Bayern: *Keine Zukunft ohne Rohstoffe. Strategien und Handlungsoptionen* (Study of the IW Consult GmbH), Munich: Vereinigung der Bayerischen Wirtschaft e.V. 2009.
23. European Commission: *Critical raw materials for the EU,* Report of the Ad-hoc-Working Group on defining critical Raw Materials, Brussels 2014. URL: http://ec.europa.eu/enter prise/policies/raw-materials/files/docs/crm-report-on-critical-raw-materials_en.pdf [accessed: 26.06.2014].
24. Buchholz, P./Huy, D./Sievers, H.: "DERA Rohstoffliste 2012, Angebotskonzentration bei Metallen und Industriemineralen – Potenzielle Preis- und Lieferrisiken", In: *DERA Rohstoffinformationen Nr. 10*, Deutsche Rohstoffagentur in der Bundesanstalt für Geowissenschaften und Rohstoffe 2012.
25. Buchholz, P./Huy, D./Liedtke, M./Schmidt, M.: *DERA-Rohstoffliste 2014, Angebotskonzentration bei mineralischen Rohstoffen und Zwischenprodukten – Potenzielle Preis- und Lieferrisiken* (DERA Rohstoffinformationen Nr. 24), Berlin, Deutsche Rohstoffagentur in der Bundesanstalt für Geowissenschaften und Rohstoffe 2015. URL: http://www.deutsche-rohstoffagentur.de/D ERA/DE/Publikationen/Schriftenreihe/schriftenreihe_node.html [Stand: accessed 04.2015].
26. US Department of Energy: *Critical Materials Strategy*, Washington DC 2011. URL: http://en ergy.gov/sites/prod/files/DOE_CMS2011_FINAL_Full.pdf [accessed 27.06.14].
27. Bast, U./Treffer, F./Thüringen, C./Elwert, T./Marscheider-Weidemann, F.: "Recycling von Komponenten und strategischen Metallen aus elektrischen Fahrantrieben". In: Thomé-Kozmiensky, K.J./Goldmann, D.: *Recycling und Rohstoffe* (Bd. 5), Neuruppin: TK Verlag Karl Thomé-Kozmiensky 2012, pp. 699–706.
28. Christlich Demokratische Union Deutschlands/Christlich-Soziale Union in Bayern/Sozialdemokratische Partei Deutschland: *Deutschlands Zukunft gestalten* (Coalition agreement between CDU, CSU und SPD, 18. legislative period. 16.12.2013), Berlin: Coalition parties of the Federal Government 2013.
29. Elsner, H./Schmidt, M./Schütte, P./Näher, U.: *Zinn – Angebot und Nachfrage bis 2020* (DERA Rohstoffinformationen Nr. 20), Deutsche Rohstoffagentur in der Bundesanstalt für Geowissenschaften und Rohstoffe 2014.
30. Liedtke, M./Schmidt, M.: "Rohstoffrisikobewertung – Wolfram", In: *DERA Rohstoffinformationen Nr. 19*, Berlin: Deutsche Rohstoffagentur in der Bundesanstalt für Geowissenschaften und Rohstoffe 2014.
31. Rosenau-Tornow, D./Buchholz, P./Riemann, A./Wagner, M.: "Assessing the long-term Supply Risks for Mineral Raw Materials – a combined Evaluation of past and future Trends". In: *Resources Policy*, 34, 2009, pp. 161–175.
32. Fettweis, G.B./Brandstätter, A./Hruschka, F.: "Was ist Lagerstättenbonität?". In: *Mitteilungen der. Österreichischen Geologischen Gesellschaft*, 78, 1985, S. 23–40.
33. US Geological Survey: *Mineral Commodity Summaries 2015*, Washington DC 2014. URL: http://minerals.usgs.gov/minerals/pubs/mcs/2015/mcs2015.pdf [accessed: 07.08.201].
34. Wellmer, F.-W.: "Reserves and Resources of the Geosphere, Terms so often misunderstood. Is the Life Index of Reserves of natural Resources a Guide to the Future?". In: *Zeitschrift der Deutschen Gesellschaft für Geowissenschaften*, 159:4, 2008, pp. 575–590.

35. Melcher, F./Buchholz, P.: "Germanium". In: Gunn, G.: *Critical Metals Handbook*, Oxford: John Wiley & Sons 2014, p. 177–203.
36. Frenzel, M./Ketris, M.P./Gutzmer, J.: "On the geological Availability of Germanium". In: *Mineralium Deposita*, 49: 4, 2014, pp. 471–487.
37. Wellmer, F.-W./Dalheimer, M./Wagner, M.: *Economic Evaluations in Exploration*, Berlin, Heidelberg, New York: Springer Verlag 2008.
38. Bradshaw, A. M./Hamacher, T.: "Nuclear fusion and the helium supply problem". In: *Fusion Engineering and Design*, 88, 2013, S. 2694–2697.
39. Rosemarin, A./Jensen, L.S.: "What is the phosphorus Challenge?", In: *European Sustainable Phosphorus Conference*, Brussels, 06.03.2013.
40. Kerr, R.A.: "The coming Copper Peak". In: *Science*, 343, 2014, pp. 722–724.
41. Wellmer, F.-W./Scholz, R.: "The Right to know the Geopotential of Minerals for ensuring Food Supply Security: The Case of Phosphorus". In: *Journal of Industrial Ecology*, 19: 1, 2015, pp. 3–6.
42. Word Wide Fund for Nature/Ecofys: *Critical Materials for the Transition to a 100% sustainable Energy Future*, WWF Report 2014, Gland, Schweiz: WWF International 2014. URL: http://www.ecofys.com/files/files/wwf-ecofys-2014-critical-materials-report.pdf [accessed: 29.06.2014].
43. Bundesanstalt für Geowissenschaften und Rohstoffe: Risikobewertung- Lithium, Hannover 2016.
44. Angerer, G./Mohring, A./Marscheider-Weidemann, F./Wietschel, M.: *Kupfer für Zukunftstechnologien – Nachfrage und Angebot unter besonderer Berücksichtigung der Elektromobilität*, Karlsruhe 2010.
45. Gordon, R.B./Bertram, M./Graedel, T. E.: "Metal Stocks and Sustainability". In: *Proceedings of the National Academy of Sciences of the USA*, 103: 5, 2006, pp. 1209–1214.
46. Tilton, J. E./Lagos, G.: "Assessing the long-run availability of copper". In: *Resources Policy*, 32, 2007, pp. 19–23.
47. Schodde, R. C.: The Key Drivers behind Resource Growth: an analysis of the Copper Industry over the last 100 Years In: Paper Mineral Economics & Management Society (MEMS) session at the 2010 SME Annual Conference, Phoenix, Arizona, March 2010), 2010. URL: http://www.minexconsulting.com/publications/Growth%20Factors%20for%20Copper%20SME-MEMS%20March%202010.pdf [accessed: 16.03.2014].
48. Schodde, R. C.: "Recent Trends in Copper Exploration — are we finding enough?". In: *International Geological Congress IGC Brisbane*, Australien, 05.–10. August 2012), 2012. URL: http://www.minexconsulting.com/publications/IGC%20Presentation%20Aug%202012%20PUBLIC.pdf [accessed: 15.01.2015].
49. Schodde, R.C.: "Global Outlook and Development Trends for Copper". In: *Philippines Mining Conference*, Manila, 20.09.2012. URL: http://www.minexconsulting.com/publications/Copper%20Outlook%20-%20PMC%20presentation%20Sept%202012.pdf [accessed: 13.03.2014].
50. Cook, P. J./Shergold, J. H.: *Phosphate Deposits of the World*, Cambridge: Cambridge University Press 1986.

Chapter 6
Conclusions

Three groups of raw materials are important for the energy systems of the future: fossil raw materials, biomass and mineral raw materials, especially the metals that are required for the construction of energy facilities. An increased demand will therefore occur during the construction of the infrastructure for a climate-neutral supply of energy.

The energy systems of the future will be significantly more diverse as compared to those of today. Wind power and photovoltaic facilities and other technologies for producing renewable energies will be added to the current systems that are dominated by coal and gas power stations. The storage technologies, such as several battery systems, storage pumping stations, hydrogen storage and compressed air storage, will be important over the long-term. On the consumer side, there will be significant technological changes such as electric vehicles, light-emitting diodes (LED) and supra-conducting magnetic heating for the processing of base metals—just to name a few. About 45 technologies are expected to be important for the conversion of the energy systems.

This technological diversity is also reflected in the demand for raw materials. The new energy technologies, as well as other high-technology products, will require increasing quantities of special metals such as the rare-earth elements and technology metals such as indium tellurium, gallium and germanium. Sixty elements, for example, are required for the manufacture of a computer chip. The energy technologies are not the only use for most of these raw materials, and there is a greater demand for many uses in the automobile industry as well as in the electronic, information and communications sectors. The energy sector is therefore in competition for the same raw materials with these other sectors.

Germany is one hundred percent depend on imports of the **primary metals**. The domestic secondary production will not be able to cover the demand in the foreseeable future, and therefore the global open commodity markets will remain critical for the future availability of raw materials in Germany.

From the geological point of view, there are sufficient metals for the implementation of the energy transition despite increasing demand for raw materials. However,

© Springer International Publishing AG, part of Springer Nature 2019
F.-W. Wellmer et al., *Raw Materials for Future Energy Supply*,
https://doi.org/10.1007/978-3-319-91229-5_6

because new mining projects require lead times of about 10–15 years from discovery to production from a new mine and lead times are also required for the expansion of existing capacity, and because nearly all raw materials are now traded on global markets, even a small increase in demand somewhere in the world can result in significant increases in prices on the global commodity markets. The tendency of concentration in the mining industry, so that increasingly more mineral occurrences belong to a decreasing number of companies in fewer producer countries, favor the formation of oligopolies and can lead to price and supply risks.

Over the long-term the real prices of nearly all raw materials have not significantly increased for about one hundred years. The effects of rationalization by technical advances in exploration, mining and processing have until now compensated for the increasingly more difficult parameters, with respect to their lower grades and increasing depths, of the mineral deposits.

However, short-term **price peaks and supply risks** could delay the implementation of the energy transition. The competition on the international commodity markets is rising fast due to the sharp increase in demand from China, and German companies must be able to assert themselves in this competitive situation. Businesses require information about critical raw materials in order to prepare themselves for supply risks and to prepare alternative strategies.

The potentially **critical raw materials** are those that have both a high supply risk and greatest economic importance. The supply risk is high if a raw material is to a large extent sourced from only a few producer countries, where the supply is classified as unreliable due to possible political crises, trade restrictions or other factors. If a raw material cannot be substituted and is only recoverable to a limited extent by recycling, this also contributes to its criticality.

Both the future technological developments in the energy sector and the availability of various raw materials on the global markets will determine which raw materials will become critical over the next years for the energy systems of the future. The technological developments in other sectors that compete with the energy sector for their raw materials will also be a significant factor. It is therefore not surprising that various studies come to different conclusions regarding the criticality of specific raw materials.

There is general agreement on the rare-earth elements, platinum group elements, indium and tellurium. These elements are classified as critical or almost critical in nearly all the recent studies on the criticality of raw materials for the energy systems of the future.

Scandium, yttrium, neodymium, dysprosium, praseodymium, terbium, europium, cerium, lanthanum and samarium belong, among others to the group of **rare-earth elements**.[1] They are required for various energy technologies such as batteries, LEDs, magnets in wind power facilities, motors and generators. The supply risks are related to the country concentration that is especially high with 95% of the global mine production in one country, China, as well as other factors. Furthermore, the recycling

---

[1]Not all the rare-earth elements are regarded as critical. There are no supply risks for samarium, for example.

rates are still insufficient because the components that contain rare-earth elements, such as magnets, are still difficult to separate as end-of-life products and there is only a limited metallurgical infrastructure for recovering these elements.

The **platinum group elements** include platinum, palladium, rhodium, ruthenium, iridium and osmium. They are important for fuel cells and catalysts in hybrid electro-vehicles and for hydrogen electrolysis, and therefore they are critical for several possible key technologies for the energy transition such as long-term storage and power-to-gas. Iridium cannot currently be substituted for certain applications (PEM-electrolyzers[2] for the production of hydrogen).

The great importance of **indium and tellurium** is related to the photovoltaics for the energy systems of the future. Indium is primarily derived from zinc production and is primarily sourced from China. The supply of tellurium is relatively diverse, with the principal supply countries in China, Japan and Belgium.

Indium, tellurium, iridium and several of the other possible critical metals are by-products, which means that they are produced as by-products during the mining of other metals. The feedback control cycle of raw material supply is only partly applicable to the **by-product metals**. A producer of the main metal, for example zinc in the case of indium, would most probably not increase production because of insufficient supply of the by-product indium. There are often only a few producers and customers, so that the market is less transparent as for other raw materials that are traded on the major exchanges. Furthermore, it is often difficult to estimate the future availability of the by-product metals.

Their good recycling characteristics is an advantage for the platinum group metals. They are used, for example, in the petro-chemical industry as catalysts, and can be recycled with good recovery and without any loss of quality. Industrial material cycles have already been established with only very minor losses occurring over the complete life-cycle. The raw material potential of the technosphere could therefore become very significant. However, it must always be considered that the recovery can only occur at the end-of-life of the products and therefore, in the case of a rapidly expanding technology such as the hydrogen electrolysis, the resources of recoverable metal in the technosphere are initially quite small and most of the demand must be covered by primary exploitation. As an alternative, less efficient technologies must be used that either require other platinum group metals or can be operated without these catalyst metals. The proportion of the demand that can be covered by **secondary production** is therefore also dependent on the timing of the conversions in the energy systems.

The classification of numerous other raw materials is not definitive. According to the assumptions about development of the future demand and the methods of assessing the criticality, some of the studies classify these raw materials as critical or almost critical, and others classify them as non-critical. This is the case for nickel, the steel alloy metal niobium, the refractory metal tungsten as well as gallium, germanium, selenium, vanadium, silver, graphite, rhenium and hafnium.

---

[2]PEM stands for "Polymer Electrolyte Membrane".

Numerous other raw materials that are required for future energy technologies are classified as being less critical. These include, among others, manganese, tantalum, molybdenum, lead, copper and lithium.

Lithium is required for batteries in electrical vehicles, and could therefore be vital for the conversion of the transport sector. Several raw material studies classify lithium as being critical, but these studies do not take the results of future exploration into account. The authors of this analysis consider that this approach is not justified because the new lithium reserves could be developed from the geopotential field by the feedback control cycle of raw material supply. A lithium shortage is therefore not expected in the next decades, although there might be occasional short-term increases in the lithium prices due to possible shortages of the supply to the market.

Some of the raw materials experts also expect a limitation of the long-term availability of copper, which is very important for the electricity sector due to its electrical conductivity. However, the authors of this analysis do not expect an imminent shortage, especially because copper, similar to the platinum group elements, can be recycled with good recoveries.

There is a certain level of criticality for **helium**, even though this element is not classified as a critical raw material in most of the criticality studies. The future energy systems may require large quantities of helium for refrigeration, and it is necessary to take the necessary measures to ensure its availability for such technological developments. Helium occurs in small quantities in natural gas deposits and, if it is not separated during the production of natural gas, then it will dissipate out of the earth's atmosphere. It can therefore not be recovered even under favorable economic conditions.

Phosphorus must be considered as critical under certain circumstances, although no supply shortages can be expected in the next decades. Phosphorus, like the other essential plant nutrients potassium and nitrogen, cannot be substituted by other substances. However, in contrast to potassium and hydrogen, the potential for phosphorus is not unlimited. There are currently approaches to focus on and monitor the international availability based on its geopotential. Today approximately half of the phosphate demand for the German agricultural industry must be imported. It is technically possible to partially recover the phosphate from sewage, but it is currently not economical.

German companies in the metal and industrial mineral sectors are increasingly using **intermediary products** that are higher in the value-added chain, and these can also be critical for the industry. There is practically no information that is available for these companies about the supply risks of these intermediary products. The German Mineral Resources Agency (DERA) at the Federal Institute for Geosciences and Natural Resources (BGR) has therefore commenced to include the important intermediary products in their criticality analyses.

Despite all these criticality studies and scenarios, the **trends in demand** and the related variations in the requirements for raw materials always remain, to a certain extent, unpredictable. The challenge for the industry is to remain flexible in their use of raw materials. By stockpiling, diversification of sources, development of substitution possibilities, and in-house as well as external recycling measures, companies could circumvent supply shortages. This means that customers are persuaded in advance to accept another material composition in their products (product clearance). The establishment of buyer groups, long-term supply contracts with price adjustment clauses, and hedging can reduce the economic risks. These efforts of the companies could be supported by governmental measures, for example the EU and WTO could take measures to remove the distortion of the competition and trade restrictions in the global commodity markets.

The **possibilities of substitution** of high risk critical materials are currently being discussed within the EU. However, some of the critical or almost critical raw materials for the energy systems of the future can only be substituted with difficulty, and these include the rare-earth elements dysprosium, yttrium, europium, lanthanum and the platinum group metals rhodium and iridium. Others, including neodymium, praseodymium and tellurium, could be replaced in several applications by other elements. Every element has its own characteristics, and therefore even for raw materials that can be readily substituted there are always some uses for which there are no known alternatives.

Supply risks and high prices of a raw material often result in an exhaustive pursuit for substitution possibilities. This can lead to the development of technologies that do not require the raw material in question. For example, for alloys formerly requiring rhenium alloys have been developed in recent years with the same performance that do not require rhenium. The effects of price movements could also be reduced by changing the use of high-temperature alloys for the production of high value steels that are required in many of the energy technologies. Sometimes complete products that depend on critical metals can be replaced by alternative technologies. For example, squirrel-cage asynchronous motors can replace synchronous motors. The advantage is that in contrast to the latter, the asynchronous motors do not contain any rare-earth elements. However, many years are often required to develop an alternative product for serial production.

The increase in the **material efficiency**, or the manufacturing of a product with smaller quantities of raw material, is another aspect of substitution. This might succeed for some technologies required for using renewable energies, but in terms of the overall economics these technologies use raw materials more intensively as compared to conventional energy facilities because the manufacture of the facilities requires more material for each unit of energy produced. Only by taking the complete life-cycle into account, from initial investment to the end of operations, is there an improvement of the raw material efficiency because renewable energies do not consume any fossil energy resources. More must be invested in the raw materials required to convert the energy systems so that there is finally an improvement in the raw material efficiency.

Recycling offers an excellent possibility to become less dependent on the critical primary raw materials. Increasingly a stock of raw materials is becoming available. It is based on old products such as automobiles, computers and infrastructure materials such as electricity transmission lines and buildings, and they comprise the so-called secondary deposits. Metals, with the same quality as from primary sources, can be recovered from secondary deposits by modern metallurgical processes. Furthermore, the lead times and investment requirements are often lower than those for primary deposits, and the social acceptance for recycling is better than it is for mining.

However, high recycling rates have only been achieved so far for the base and precious metals. One reason for this is the lack of proven recycling processes for some of the potentially critical raw materials such as the rare-earth elements or indium, germanium, gallium or tantalum that occur in complex associations. In recycling processes focusing on valuable metals such as copper or precious metals, these minor metals separate into the slags and are therefore lost.

The **collection efficiency** is often low, and presents another problem. Although some very efficient recycling process chains have become established in industrial applications, for example the platinum group elements in process catalysts, there is often a lack of an economic incentive to deliver consumer products for recycling at the end of their lives. The political and social situation, particularly the waste regulations as well as their enforcement, are vital.

Even if recycling can be expanded for many raw materials, one hundred percent recovery is not possible because of thermodynamics constraints, and also not sensible with respect to the energy requirements. The recovery becomes increasingly inefficient and more energy intensive depending on the complexity of the compounds that contain the raw materials. The **product design** can contribute to improving the possibilities for recycling, for example if components with valuable raw materials could be easily removed from products.

Recycling can contribute significantly to the supply of important metals, but on its own it cannot cover the demand. The conversion of the energy systems therefore remains dependent on mining, also because of the initial increase in the requirements for metals.

In the future, **innovation in exploration and mining** is necessary for discovering and exploiting new primary deposits. Initially the near-surface mineral deposits were exploited, so that the potential for new discoveries is at greater depth. The discovery of these ore deposits is made possible by, among others, the new developments in electromagnetic exploration technology. In the longer term, marine natural resources may be significant for the exploitation of copper, nickel, cobalt and various other high-technology metals from polymetallic nodules (manganese nodules), cobalt-rich manganese crusts and massive sulfides. However, very significant research and development on their exploitation and processing is still required.

New technological developments are also involved in the discovery of new deposits of the fossil fuel natural resources. In particular, **shale gas and shale oil** should be mentioned as they can be recovered by hydraulic-fracturing technology (fracking), which is the subject of a very controversial public debate. The currently low prices for crude oil and natural gas on the global markets are the result of the increased production of shale oil and shale gas in the USA. In the long-term, deep marine methane clathrates could also be significant.

Even if the production from the unconventional deposits should increase appreciably, **crude oil and natural gas** remain as raw materials with limited resources. In the long-term, high and even increasing prices for these energy fuels are therefore to be expected despite the current fall in prices. Germany is very dependent on imports of both crude oil and natural gas from just a few producing countries. Domestic sources only cover 12% of the German consumption of natural gas and two percent of crude oil.

The dependency on Russia for natural gas could be reduced by the increased use of liquefied natural gas (LNG) that can be transported by tanker from other source countries. Another alternative is to produce natural gas from unconventional deposits in Germany by means of fracking.

Even though the electricity generation from wind power and photovoltaic continues to be expanded, and therefore reducing the demand for fossil fuels in the long-term, either coal or natural gas power stations must be kept ready as reserves in the foreseeable future, so long as no long-term storage of electricity is available. Natural gas power stations have the advantage of being more flexible than the coal power stations, and can therefore better compensate the fluctuating supply from wind power and photovoltaic. Furthermore, the combustion of natural gas is much cleaner than coal and causes less CO_2 emissions. At the same time, coal is much less expensive, is available over the long-term and, at least in the case of lignite, can be exploited from domestic deposits.

Whereas shortages of the metallic raw materials are primarily related to developments on the commodity markets, the availability of **bioenergy** is constrained by physical limits. The estimates of how much bioenergy will be available globally from agrarian biomass in 2050 are wide apart—from 50 to 500 exajoule per year. In addition to the increasing demand for foodstuffs, if the environmental impacts from intensive agriculture—greenhouse gas emissions, loss of biodiversity, increased usage of water, contamination of water and soil degradation—are also taken into consideration, the authors believe the available quantities of agrarian bioenergy cannot be significantly increased. Measures to ensure the supply of bioenergy should be focused on the demand, for example by using biomass in all sectors as efficiently as possible, and using bioenergy only in such sectors where the overall effect on the system is very positive.

Agrarian biomass waste represents a significant potential for energy production. More efficient manufacturing chains for foodstuffs—currently about sixty percent of the global harvest is lost in the agriculture and is also lost in the supply chains as well as being thrown out by supermarkets and households. A nutritional diet with less meat products could also free up some of the agricultural areas that could then possibly be used to produce bioenergy.

Bioenergy has a lower area efficiency as compared to photovoltaic and wind power generators, and a worse greenhouse gas balance with the exception of wood from sustainably managed forests. However, because of its storage capability and high energy density, bioenergy can assume some of the functions in the energy system for which wind and solar energy are less suitable, for example the bridging of long calm periods and as a fuel for heavy goods traffic.

The availability of **water** is a constraining factor for the agricultural productivity in many regions. Currently about seventy percent of the global consumption of freshwater is used for irrigation. Water, as salt water, is available in practically unlimited quantities and can be converted into freshwater by the energy-intensive process of desalination. The water issue therefore changes to become an issue of energy availability.

Water availability is considered by some experts also to be a limiting factor for mining, since many mining districts are located in arid or semi-arid regions. The use of freshwater can be partly reduced by using brackish or salty water. Desalinated sea water is also used in mining.

The environmental balance of mineral exploitation is also of importance for the **energy requirements**. In future, more energy will probably be required per tonne of metal produced because of the increasing depths of the mineral deposits that must be mined and the more complex ores that must be processed. The higher CO_2 emissions related to this mining could be compensated if the mines use energy specifically from renewable sources. It is even possible that the mining companies use the less expensive excess wind and solar electricity to process ores with especially low metal grades. In this way the companies could use electricity more flexibly, and contribute to the balancing of fluctuating inputs.

In addition to the technical and economic requirements for a mining project, there is one other important issue: **the social acceptance**, which implies that the people must support, or at least tolerate, mining. Winning, or maintaining, this social acceptance for mineral exploitation, the so-called "social license to operate" is an increasingly important challenge for the mining industry. The attitudes of the people towards mining in their own country, either acceptance or rejection, depends on a number of factors: the level of development of the country and its economic dependency on exploitation of its natural resources; foreign exchange and tax income; jobs; and the industrial infrastructure that is generated by mining. Sustainable and socio-ecologically acceptable exploitation of natural mineral resources can only be established if the various interests are discussed and evaluated. Social acceptance for mining will only be finally attained if the local population can be convinced that their values are respected, the environmental impacts are minimized, and the economic advantages are reflected in jobs and improved infrastructure.

Because of their stronger impact on the landscape and resettlement of people, **open pit** projects often receive more criticism than underground projects. The trend away from the increasingly large open pit operations, could contribute to maintaining or even increasing the social acceptance. However underground mining is also more costly.

In many countries outside Europe, mining has often resulted in serious consequences for human health, the environment and social structures. Illegal and dubious exports of electronic scrap and end-of-life products in regions with inadequate recycling standards are also a problem for the recycling industry. The establishment of higher **environmental and social standards** throughout the world will be a major challenge for the future of the raw materials industry, if not the single most important challenge.

The natural resource industry can contribute itself to the solution of this problem by cooperation among its companies to define binding standards. Although this is already the practice of the major international mining companies, many of the small and medium-size mining companies often do not operate to these standards and cause a disproportionate amount of environmental damage in relation to their production of mineral resources. The international banks can also demand these standards as a precondition for the financing of mining projects, and therefore they also have a special socio-ecological responsibility.

Appendix A

See Fig. A.1.

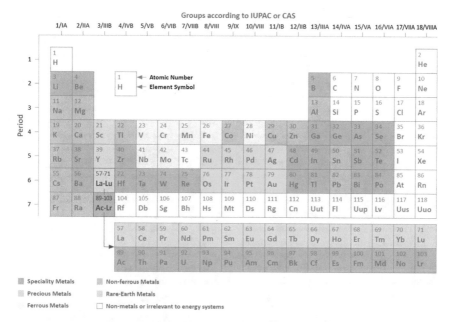

Groups according to IUPAC or CAS

Fig. A.1 Distribution of elements according to the types of metal (Modified according to UNEP [1, p. 11]). It should be noted that many of the commonly used terms such as technology metals, high technology metals, electronic metals, special metals, refractory metals etc. or the exact chemical classifications have not been considered since they often overlap with each other (with permission of the United Nations Environment Program UNEP)

© Springer International Publishing AG, part of Springer Nature 2019
F.-W. Wellmer et al., *Raw Materials for Future Energy Supply*,
https://doi.org/10.1007/978-3-319-91229-5

See Fig. A.2.

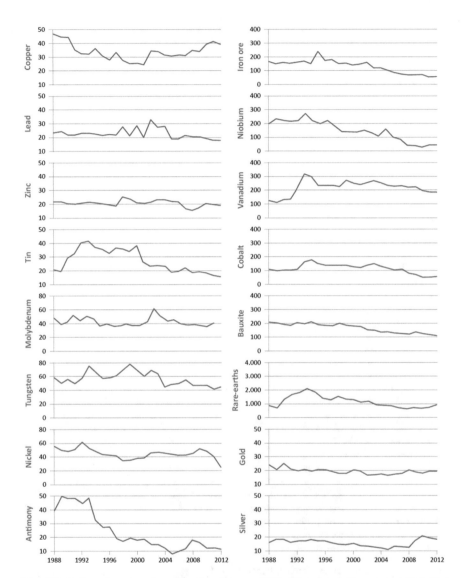

Fig. A.2 Ratios of reserves to production (in years) for selected metallic raw materials that are most relevant for energy systems of the future [2]. For most of these raw materials, this value has decreased during the past years, which is primarily due to the increased demand from China. Exploration successes always occur sometime later. A decrease indicates a potential shortage of these raw materials. In some cases supply criticality can be deduced considering an average ten years for the development for new mining projects

See Table A.1.

Table A.1 Energy technologies analyzed in the DOE, JRC-IET and Wuppertal Institute for their relevance to raw materials consumption ([2, 3, 4])

Study	DOE 2011	JRC-IET 2013	WI 2014 WI-1	WI-2
Centralized production technologies				
Coal-fired power plant		X		
Gas-fired power plant		X		
Combined-cycle power plant		X		
Integrated gasification combined cycle (IGCC) power plant		X		
Carbon capture and storage (CCS)		X		
Decentralized production technologies				
Combined heating power plant (CHP) with combustion engine		X		
Combined heating power plant (CHP) with Stirling engine		X		
Proton exchange membrane fuel cell (PEM)		X		
Solid oxide fuel cell (SOFC)		X	X	
Decentralized production technologies				
Wind power generator with gear box	X	X	X	
Direct drive wind power generator	X	X	X	
Hydro power plant (Kaplan, Francis, Pelton)		X		X
Tidal power plant		X		
Geothermal power plant		X		X
Silicon solar cells (Photovoltaic PV)	X	X		X
Cadmium telluride (CdTe) solar cells (PV)	X	X	X	
Copper indium diselenide solar cells (PV)	X	X	X	
Copper indium gallium diselenide solar cells (PV)	X	X	X	
Gallium arsenide solar cells (PV)		X		
Concentrating solar power (CSP)		X	X	
Thermal solar power		X	X	
Storage technologies				
Pump storage hydropower		X		X
Compressed air energy storage (CAES)		X		X
Hydrogen storage		X		X
Lead-acid battery		X		
Nickel-cadmium battery	X	X		
Nickel-metal hydride battery	X			
Lithium-ion battery	X	X		X
Sodium-sulfur battery		X		X
Redox flow battery		X	X	

(continued)

Table A.1 (continued)

Study	DOE 2011	JRC-IET 2013	WI 2014 WI-1	WI-2
Power grids				
Overhead power line (aluminum)				
Underground cable (copper)				X
Electrical mobility				
Hybrid vehicles	X	X	X	
Battery electric vehicles	X	X	X	
Fuel cell vehicles		X	X	
Flywheel energy storage		X		
Supercapacitor electricity storage		X		
Exploitation of energy sources				
Biomass to liquid synthetic fuel (BtL)				X
Bioethanol fermentation				X
Diaphragm biogas refining to natural gas				X
Power to gas electrolysis			X	
Efficiency technologies—range				
Fluorescent lamps	X	X		
LED (light emitting diode) lamps	X	X		
Synchronous motors (permanent magnetic excitation)	X			
Efficiency technologies—industrial (examples)				
Superconducting magnetic heat process in nonferrous metal processing		X		

The technologies mentioned by the Wuppertal Institute are subdivided in WI-1: technologies that have been analyzed in depth with respect to potentially critical raw materials; and WI-2: technologies that have been evaluated with respect to potentially critical raw materials, but have not been analyzed in depth since no relevance for critical raw materials was identified

References

1. United Nations Environment Programme: *Recycling Rates of Metals – a Status Report*, Report of the Working Group on the Global Metal Flows to the International Resource Panel, 2011. URL: http://www.unep.org/publications/ [accessed 15.01.2015].
2. Bundesanstalt für Geowissenschaften und Rohstoffe: *BGR-Datenbank*, Hannover: Bundesanstalt für Geowissenschaften und Rohstoffe 2014.
3. Moss, R. L./Tzimas, E./Willis, P./Arendorf, J./Tercero Espinoza, L.: *Critical Metals in the Path towards Decarbonisation of the EU Energy Sector – Assessing Rare Metals as Supply-Chain Bottlenecks in Low-Carbon Energy Technologies* (Scientific and Policy Reports), Petten: European Commission, Joint Research Centre, Institute for Energy and Transport 2013. URL: https://setis.ec.europa.eu/newsroom-items-folder/new-jrc-report-critical-metals-energy-sector [accessed: 02.05.2014].
4. US Department of Energy: *Critical Materials Strategy*, Washington DC 2010. URL: http://www.energy.gov/sites/prod/files/piprod/documents/cms_dec_17_full_web.pdf [accessed: 28.10.2014].
5. Wuppertal Institut für Klima, Umwelt, Energie GmbH: *KRESSE — Kritische mineralische Rohstoffe bei der Transformation des deutschen Energieversorgungssystems*, Abschlussbericht an das Bundesministeriums für Wirtschaft und Energie, Wuppertal 2014. URL: http://wupperinst.org/de/projekte/details/wi/p/s/pd/38/ [accessed: 15.12.2014].

Appendix B
Examples of Quantification of the Raw Material Requirements for Individual Energy Technologies Based on the Studies by the Wuppertal Institute and JRC-IET [1, 2]

See Table B.1.

Table B.1 Specificrequirements for potentially criticalmineral raw materials for the constructionof wind powergenerators [1] (with permission of Wuppertal Institute)

Type of generator	Raw Material	Present time	2025	2050
	(kg/MW)			
Synchronous motor, permanent field				
Direct drive (DD)	Neodymium	201.5	162.5	130.0
	Dysprosium	15.0	11.7	11.7
Medium speed drive (MS)	Neodymium	49.6	40.0	32.0
	Dysprosium	3.7	2.9	2.9
High speed drive (HS)	Neodymium	24.8	20.0	16.0
	Dysprosium	1.8	1.4	1.4
Synchronous motor, electric field				
High temperature super conductor	Yttrium	–	2.3	2.3

See Tables B.2, B.3, B.4, B.5 and Figs. B.1, B.2, B.3.

© Springer International Publishing AG, part of Springer Nature 2019
F.-W. Wellmer et al., *Raw Materials for Future Energy Supply*,
https://doi.org/10.1007/978-3-319-91229-5

Table B.2 Specific requirements for potentially critical mineral raw materials for current and future photo-voltaic facilities (From Wuppertal Institute [1, p. 156ff]; modified after Schlegl [3]) (with permission of Wuppertal Institute)

Element	Symbol	Polycrystalline silicon (c-Si)	Amorphous silicon (a-Si)			Cadmium-telluride (CdTe)			Copper-indium-gallium diselenide (CI(G)S)		
		(kg/MW$_p$)									
		2013	2013	2025	2050	2013	2025	2050	2013	2025	2050
Silicon	Si	*6000*	37			–			–		
Silver	Ag	*62*				–			–		
Copper	Cu	*630*	NA			*206*			*21*		
Cadmium	Cd	–	–			*116.7–143*	63.8	33.0	Up to 1.3	1.3	0
Tellurium	Te	–	–			*99.7–135*	43.1	35.3	–		
Indium	In	–	*4.0 –5*	0	0	Up to 15.5	15.5	0	*55.5–75*	45.0	3.0
Gallium	Ga	–	–						*2–7.2*	3.2	1.2
Selenium	Se	–	–						*10–39.3*	17.4	6.3
Germanium	Ge	–	–			–			–		

Comparison of the current specific material requirement in kilogram per megawatt of electricity produced from photovoltaic facilities as well as an estimate of the requirement for raw materials regarded as critical (bold). In the sources used by the KRESSE study (italics = values from Schlegl) there are some different estimates for the current requirements for raw materials, so that only a range of values for some potentially critical raw materials is provided in this summary. The estimation of the future indium requirement excludes indium oxide mixed with tin (ITO, indium tin oxide) because of the expected technical developments and the resultant material substitution

Table B.3 Specific requirements for potentially critical mineral raw materials for geothermal power stations and fuel cell technologies (After Moss et al. [2, pp. 23, 31]) (with permission of JRC-IET)

Geothermal power plant			Fuel cells		
Element	Symbol	Requirement (kg/MW)	Element	Symbol	Requirement (kg/MW)
Tantalum	Ta	64	Platinum	Pt	67.9
Nickel	Ni	120,155	Ruthenium	Ru	22.0
Molybdenum	Mo	7209	Chromium	Cr	48,323.2
Chromium	Cr	64,405	Nickel	Ni	282.0
Niobium	Nb	128	Yttrium	Y	8.4
Titanium	Ti	1634	Lanthanum	La	38.0
Copper	Cu	3605	Gadolinium	Gd	1.2
Manganese	Mn	4325	Samarium	Sm	1.2
			Cerium	Ce	8.0
			Cobalt	Co	10.8

The specific requirements for steel alloy elements for EGS power stations (Enhanced Geothermal Systems (Enhanced Geothermal System is the technical term for deep geothermal facilities to exploit underground energy, by which various technical measures are used to improve the rock permeability for the heat carrier medium (usually water), for example by hydraulic stimulation (fracking))) with a capacity of 50 MW_{el} based on 25 drill holes each to a depth of 5 km (From Moss et al. [2]; compare Wuppertal Institute, p. 83), or on the example of fuel cells: specific material demand in kilogram per megawatt capacity

Table B.4 Specific requirements of potentially critical mineral raw materials for electro-vehicles (excluding battery systems) (After Moss et al. [2, p. 56]) (with permission of JRC-IET)

Type of motor	Metal/material	HEV (2012)	HEV (2020–)	PHEV	BEV	FCV	Mild hybrid
Permanent magnet	Neodymium (Nd)	0.76	0.91	1.46	2.55	2.92	0.36
	Iron (Fe)	2.25	2.25	3.6	7.65	8.1	0.90
	Dysprosium (Dy)	0.29	0.14	0.22	0.38	0.38	0.05
	Boron (B)	0.025	0.025	0.04	0.085	0.09	0.01
Induction	Copper (Cu)	25	25	40	70	80	10

HEV = hybride electrical vehicles, PHEV = plug-in electrical vehicles, BEV = battery electrical vehicles, FCV = fuel cell vehicles
Metallic raw material requirements in kilogram per vehicle

Table B.5 Specific requirements of potentially critical mineral raw materials for battery systems in electro-vehicles (After Moss et al. [2, p. 52ff]) (with permission of JRC-IET)

Battery system	Metal/material	BEV	HEV	PHEV-90	FCV
Nickel-metal hydride (NiMH)	Nickel (Ni)	–	6.67	–	–
	Cobalt (Co)	–	1.16	–	–
	Lanthanum (La)	–	1.16	–	–
	Cerium (Ce)	–	0.77	–	–
	Neodymium (Nd)	–	0.23	–	–
	Praseodymium (Pr)	–	0.08	–	–
	Samarium (Sm)	–	0.08	–	–
Nickel-cobalt-aluminum (NCA)	Copper (Cu)	6.23	0.13	1.06	0.17
	Cobalt (Co)	46.65	0.97	7.97	1.26
	Lithium (Li)	8.44	0.18	1.44	0.23
	Aluminum (Al)	1.35	0.03	0.23	0.04
	Graphite	72.19	1.50	12.33	1.96
	Copper (Cu; anode)	66.82	1.39	11.41	1.81
Lithium-iron phosphate (LFP)	Lithium (Li)	4.52	0.09	0.77	0.12
	Iron (Fe)	39.08	0.81	6.68	1.06
	Graphite	85.54	1.79	14.68	2.33
	Copper (Cu; anode)	79.54	1.66	13.59	2.15
Lithium-manganese spinel (LMS)	Lithium (Li)	7.81	0.16	1.33	0.21
	Manganese (Mn)	60.07	1.25	10.26	1.63
	Copper (Cu; anode)	71.08	1.48	12.14	1.93
	Titanium (Ti)	38.78	0.81	6.63	1.05
Nickel-cobalt-manganese (NCM)	Lithium (Li)	4.64	0.10	0.79	0.13
	Nickel (Ni)	14.43	0.30	2.46	0.39
	Cobalt (Co)	13.91	0.29	2.38	0.38
	Manganese (Mn)	12.88	0.27	2.20	0.35
	Graphite	53.08	1.11	9.07	1.44
	Copper (Cu; anode)	49.13	1.02	8.39	1.33

Metallic raw material requirements in kilogram per vehicle

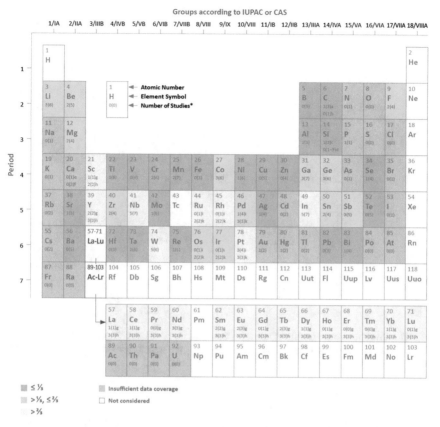

Fig. B.1 Erdmann and GraedelStudy (modified from Erdmann and Graedel [4, p. 7625])—**comparisonof seven criticality studies**. As compared to Fig. B.2, the elements are presented here in part only as the mineral occurrence of the raw material or in the context of a metal group (see the relevant labelling a to k; this is particularly relevant for the rare-earth elements and the platinum group metals) . In some studies, mineral raw materials or metal groups were analyzed instead of each individual element according to the following system: Carbon: a—diamond, b—graphite; silicon: c—elemental, d—silicates and clay; calcium: e—limestone, f—gypsum; rare-earth elements: g—elemental, h—as a member of the rare-earth element group; platinum group metals (PGM): i—elemental, k—as a member of the platinum group metals (with permission of the American Chemical Society)

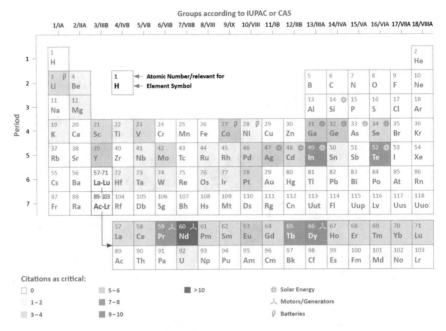

Fig. B.2 Meta-analysis by the Wuppertal Institute [1, p. 48]. A comparison of twelve studies on potentially critical raw materials that are relevant to the transformation of the German energy supply system (with permission of the Wuppertal Institute)

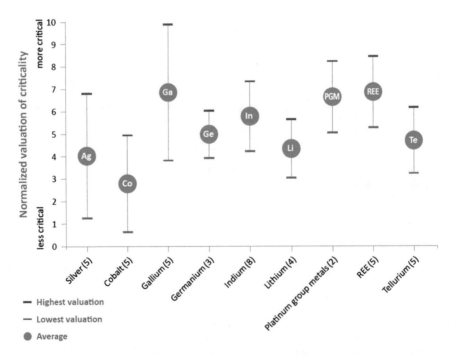

Fig. B.3 Meta-analysis by the UK Energy Research Centre [5]. Normalized criticality ranges for 9 elements were derived from 11 studies on low-carbon energy technologies. The values in brackets indicate the number of criticality studies that assessed the respective elements as critical (with permission of the UK Energy Research Centre)

References

1. Wuppertal Institut für Klima, Umwelt, Energie GmbH: *KRESSE—Kritische mineralische Rohstoffe bei der Transformation des deutschen Energieversorgungssystems*, Abschlussbericht an das Bundesministeriums für Wirtschaft und Energie, Wuppertal 2014. URL: http://wupperinst.org/de/projekte/details/wi/p/s/pd/38/ [accessed: 15.12.2014].
2. Moss, R. L./Tzimas, E./Willis, P./Arendorf, J./Tercero Espinoza, L.: *Critical Metals in the Path towards Decarbonisation of the EU Energy Sector – Assessing Rare Metals as Supply-Chain Bottlenecks in Low-Carbon Energy Technologies* (Scientific and Policy Reports), Petten: European Commission, Joint Research Centre, Institute for Energy and Transport 2013. URL: https://setis.ec.europa.eu/newsroom-items-folder/new-jrc-report-critical-metals-energy-sector [accessed: 02.05.2014].
3. Schlegl, T.: "Entwicklungslinien der PV-Technologie und Materialsubstitutionsmöglichkeiten In: *Tagung "Strategische Metalle für die Energiewende"*, Tutzing 2013.
4. Erdmann, L./Graedel, T.E.: "Criticality of non-fuel minerals: a review of major approaches and analyses". In: *Environmental Science and Technology*, Bd. 45: 18, 2011, pp. 7620–7630.
5. United Kingdom Energy Research Centre (UKERC): *Materials Availability. Comparison of Material Criticality Studies – Methodologies and Results* (Working Paper III), 2013.

Appendix C
Lists of the studies that were evaluated by the KRESSE study,[1] Erdmann and Graedel[2] and the UK Energy Research Centre[3] for Their Comparative Analyses

(Underlined are those studies that were used in at least two of the reviewed meta-studies; because of different citation formats, the title and the reference for the Kresse study are listed according to the citation in that study).

(1) KRESSE-study of the Wuppertal Institute (2014).

• Materials critical to the energy industry—An introduction (Achzet et al. 2011)

Achzet, B./Reller, A./Zepf, V./Rennie, C./Ashfield, M./Simmons, J.: *Materials critical to the Energy Industry. An introduction* (Report for the BP Energy Sustainability Challenge), University Augsburg 2011. URL: http://www.physik. uni-augsburg.de/lehrstuehle/rst/downloads/Materials_Handbook_Rev_2012.pdf [accessed: 28.10.2014].

• Energy Critical Elements: Securing Materials for Emerging Technologies (American Physical Society 2011)

American Physical Society/Material Research Society (APS/MRS): *Securing Materials for Emerging Technologies* (Report of the APS Panel on public affairs und der MRS), Washington DC 2011.

• Rohstoffe für Zukunftstechnologien (Angerer et al. 2009)

Angerer, G./Erdmann, L./Marscheider-Weidemann, F./Lullmann, A./Scharp, M./ Handke, V./Marwede, M.:*Raw Materials for emerging Technologies* (Englische summary of the report of Fraunhofer ISI (Institut für System- und Innovationsforschung) und des IZT (Institut für Zukunftsstudien und Technologiebewertung) for the German Federal Ministry for Economics and

---

[1]Wuppertal Institute 2014.
[2]Erdmann and Graedel 2011.
[3]UKERC 2013.

© Springer International Publishing AG, part of Springer Nature 2019
F.-W. Wellmer et al., *Raw Materials for Future Energy Supply*,
https://doi.org/10.1007/978-3-319-91229-5

Technology, Stuttgart: Fraunhofer IRB Verlag 2009. URL: http://www.isi.
fraunhofer.de/isi-en/service/presseinfos/2009/pri09-02.php [Stand: 27.10.2014].

Angerer, G./Erdmann, L./Marscheider-Weidemann, F./Lullmann, A./Scharp,
M./Handke, V./Marwede, M.:*Rohstoffe für Zukunftstechnologien: Einfluss des
branchenspezifischen Rohstoffbedarfs in rohstoffintensiven Zukunftstechnologien
auf die zukünftige Rohstoffnachfrage* (Report of the Fraunhofer ISI (Institut für
System- und Innovationsforschung) und IZT (Institut für Zukunftsstudien und
Technologiebewertung) for the German Federal Ministry for Economics and
Technology, Stuttgart: Fraunhofer IRB Verlag 2009. URL: http://www.isi.
fraunhofer.de/isi-en/service/presseinfos/2009/pri09-02.php [Stand: 27.10.2014].

- Risk List 2012 (BGS 2014)

British Geological Survey (BGS): *Risk List 2012 – Current supply risk index for
chemical elements or element groups which are of economic value*, Nottingham,
UK.: British Geological Survey 2012. URL: http://www.bgs.ac.uk/mineralsuk/
statistics/risklist.html [accessed: 04.12.2014].

- Critical Materials Strategy (U.S. DOE 2010)

US Department of Energy (DOE): *Critical Materials Strategy,* Washington DC
2010. URL: http://www.energy.gov/sites/prod/files/piprod/documents/cms_dec_
17_full_web.pdf [accessed: 28.10.2014].

- Critical raw materials for the EU (European Commission 2010)

European Commission (EC): *Critical raw materials for the EU* (Report of the
Ad-hoc-Working Group on defining critical Raw Materials), Brussels 2010. URL:
http://ec.europa.eu/enterprise/policies/raw-materials/files/docs/report-b_en.pdf [ac-
cessed: 01.05.2014].

- Future Metal Demand from Photovoltaic Cells and Wind Turbines,
  Investigating the Potential Risk of Diasabling a Shift to Renewable Energy
  Systems (European Parliament 2011)

European Parliament 2011: Future Metal Demand from Photovoltaic Cells and
Wind Turbines Investigating the Potential Risk of Disabling a Shift to Renewable
Energy Systems, Brussels: Science and Technology Options Assessment (STOA)
2011. URL: http://www.europarl.europa.eu/RegData/etudes/etudes/join/2011/
471604/IPOLJOIN_ET%282011%29471604_EN.pdf [accessed: 13.05.2014; currently not valid anymore].

- Trends der Angebots- und Nachfragesituation bei mineralischen Rohstoffen
  (Frondel et al. 2006)

Frondel, M./Grösche, P./Huchtemann, D./Oberheitmann, A./Peters, J./Angerer, G./
Sartorius, C./Buchholz, P./Röhling, S./Wagner, M.: *Trends der Angebots- und
Nachfragesituation bei mineralischen Rohstoffen*, (Endbericht an das BMWi,
Forschungsprojekt Nr.09/05), Bundesanstalt für Geowissenschaften und Rohstoffe

(BGR)/ Fraunhofer Institut für System- und Innovationsforschung (Fraunhofer ISI)/ Rheinisch-Westfälisches Institut für Wirtschaftsforschung (RWI) 2006.

- Critical Metals in Strategic Energy Technologies – Assessing Rare Metals as Supply-Chain Bottlenecks in Low-Carbon Energy Technologies (Moss et al. 2011)

Moss, R.L./Tzimas, E./Kara, H./Willis, P./Kooroshy, J.:*Critical Metals in Strategic Energy Technologies* (Report EUR 24884 EN), Petten, Niederlande: European Commission, Joint Research Centre, Institute for Energy and Transport (JRC-IET) : 2011.

National Research Council of the National Academies (NRC):*Minerals, critical minerals, and the U.S. economy. Prepublication Version*, Washington D.C.: The National Academies Press 2007. URL: http://www.nma.org/pdf/101606_nrc_study. pdf [accessed: 27.10.2014].

- Seltene Erden – Daten & Fakten (Schüler 2011)

Schüler, D.: *Seltene Erden – Daten & Fakten* (Background Paper), Berlin: Öko-Institut e.V. 2011.

- Rohstoffkonflikte nachhaltig vermeiden: Rohstoffe zwischen Angebot und Nachfrage (Supersberger und Ritthoff 2010)

Supersberger, N./Ritthoff, M.: *Rohstoffkonflikte nachhaltig vermeiden: Rohstoffe zwischen Angebot und Nachfrage - Teilbericht 2* (Study of the Wuppertal Instituts für Klima, Umwelt, Energie und adelphi for the Umweltbundesamt, FKZ 370819 102), Wuppertal 2010.

(2) **Erdmann und Graedel (2011) [Underlined are the studies that that were used also in the KRESSE-study and also by the UK Energy Research Centre (2013)].**

Bae, J.-C.: "Strategies and Perspectives for Securing Rare Metals in Korea". In: *Critical Elements for New Energy Technologies* (Proceedings of the Energy Initiative Workshops, 29. April 2010), Cambridge, MA: Massachusetts Institute of Technology (MIT) 2010. URL: http://web.mit.edu/miteicomm/web/reports/critical_ elements/CritElem_Report_Final.pdf [accessed: 04.12.2014].

European Commission (EC): *Critical raw materials for the EU* (Report of the Ad-hoc-Working Group on defining critical Raw Materials), Brussels 2010. URL: http://ec.europa.eu/enterprise/policies/raw-materials/files/docs/report-b_en.pdf [accessed: 01.05.2014].

Halada, K./Shimida, M./Ijima, K.: "Forecasting of the Consumption of Metals up to 2050". In: *Mater. Trans.*, 49: 3, 2008, S. 402–410. URL: http://dx.doi.org/10. 2320/matertrans.ML200704.

Morley, N./Eatherley, D.: *Material Security: Ensuring resource availability to the UK economy*, Chester, UK: Resource Efficiency KTN/Oakdene Hollins/C-Tech Innovation 2008.

Shinko Research (Mitsubishi UFJ Research and Consulting): *Trend Report of Development in Materials for Substitution of Scarce Metals* (Report 08007835-0080078380), Tokyo: New Energy and Industrial Technology Development Organization (NEDO) 2009.

National Research Council of the National Academies (NRC): *Minerals, Critical Minerals, and the U.S. Economy*, Washington, D.C.: The National Academies Press 2008.

Pfleger, P./Lichtblau, K./Bardt, H./Reller, A.: *Rohstoffsituation Bayern: Keine Zukunft ohne Rohstoffe. Strategien und Handlungsoptionen* (Study of the IW Consult GmbH), Munich: Vereinigung der Bayerischen Wirtschaft e.V. 2009.

(3) **UK Energy Research Centre (2013) [Underlined are the studies that that were used also in the KRESSE-study and also by Erdmann und Graedel (2011)].**

AEA Technology Plc: *Review of the Future Resource Risks Faced by UK Business and an Assessment of Future Viability* (Department for Environment, Food and Rural Affairs, Defra) London, UK: Department for Environment, Food and Rural Affairs (Defra) 2010.

Angerer, G./Erdmann, L./Marscheider-Weidemann, F./Lullmann, A./Scharp, M./Handke, V./Marwede, M.:*Raw Materials for emerging Technologies* (Englische summary of the report of Fraunhofer ISI (Institut für System- und Innovationsforschung) und des IZT (Institut für Zukunftsstudien und Technologiebewertung)for the German Federal Ministry for Economics and Technology), Stuttgart: Fraunhofer IRB Verlag 2009. URL: http://www.isi.fraunhofer.de/isi-en/service/presseinfos/2009/pri09-02.php [accessed: 27.10.2014].

Angerer, G./Erdmann, L./Marscheider-Weidemann, F./Lullmann, A./Scharp, M./Handke, V./Marwede, M.: *Rohstoffe für Zukunftstechnologien: Einfluss des branchenspezifischen Rohstoffbedarfs in rohstoffintensiven Zukunftstechnologien auf die zukünftige Rohstoffnachfrage* (Report of the Fraunhofer ISI (Institut für System- und Innovationsforschung) und IZT (Institut für Zukunftsstudien und Technologiebewertung) for the German Federal Ministry for Economics and Technology), Stuttgart: Fraunhofer IRB Verlag, 2009. URL: http://www.isi.fraunhofer.de/isi-en/service/presseinfos/2009/pri09-02.php [accessed: 27.10.2014].

British Geological Survey (BGS): *Risk List 2011*, Nottingham, UK: BGS 2011.

US Department of Energy (DOE): *Critical Materials Strategy*, Washington DC 2010. URL: http://www.energy.gov/sites/prod/files/piprod/documents/cms_dec_17_full_web.pdf [accessed 28.10.2014].

US Department of Energy (DOE): *Critical Materials Strategy*, Washington DC 2011. URL: http://energy.gov/sites/prod/files/DOE_CMS2011_FINAL_Full.pdf [accessed 27.06.14].

European Commission (EC): *Critical raw materials for the EU* (Report of the Ad-hoc-Working Group on defining critical Raw Materials), Brussels 2010. URL: http://ec.europa.eu/enterprise/policies/raw-materials/files/docs/report-b_en.pdf [accessed: 01.05.2014].

Graedel, T.E./Barr, R./Chandler, C./Chase, T./Choi, J./Christoffersen, L./ Friedlander, E./Henly, C./ Jun, C./Nassar, N.T./Schechner, D./Warren, S./Yang, M.-Y./Zhu, C.: "Methodology of Metal Criticality Determination". In: Environmental Science & Technology, 46: 2, 2012, S. 1063–1070.

Moss, R.L./Tzimas, E./Kara, H./Willis, P./Kooroshy, J.:*Critical Metals in Strategic Energy Technologies* (Report EUR 24884 EN), Petten, Niederlande: European Commission, Joint Research Centre, Institute for Energy and Transport (JRC-IET) : 2011.

Morley, N./Eatherley, D.: *Material Security: Ensuring resource availability to the UK economy*, Chester, UK: Resource Efficiency KTN/Oakdene Hollins/C-Tech Innovation 2008.

National Research Council of the National Academies (NRC):*Minerals, critical minerals, and the U.S. economy. Prepublication Version*, Washington D.C.: The National Academies Press 2007. URL: http://www.nma.org/pdf/101606_nrc_study. pdf [accessed: 27.10.2014].

Scottish Environmental Protection Agency (SEPA): *Raw materials critical to the Scottish economy* (Report by AEA Technology for the SEPA and the Scotland and Northern Ireland Forum For Environmental Research, SNIFFER), Edinburgh, Scotland 2011.

Glossary

Backward integration A company exhibits backward integration if it acquires one or more of the upstream production steps (changing the input situation of the company): the company therefore itself then produces the materials that were previously purchased, and subsequently processed further by the company.

By-product elements Occurrences of by-product elements are associated with another primary commodity. The oremineral phases of the primary commodity and the by-product may be closely intergrown, or the by-product element is integrated into the crystal lattice of the primary commodity. The different commodities can often only be separated with significant use of energy. The recovery of the by-product elements is therefore unavoidably coupled with the recovery of the primary commodity, and the by-products are therefore distinguished from commodities that form their own mineral deposits.

Criticality "Critical" raw materials and raw materials of "strategic economic importance" are terms often used with respect to the availability of these raw materials. The term "critical" does not relate to the raw material itself or to the size of the reserves or resources, but reflects the availability of the raw material and its importance for the economy. The source countries and their political stability, as well as the supplyconcentration, are important factors in determining criticality. The environmental impact caused by the exploitation of the resources is often also included in the evaluation. The term "criticality" has become established as a reference to the availability of raw materials. It should here be noted that the term criticality related to the availability of natural resources does not have any relationship with the classical definition of the term in physics.

Cut-off grade Most metalmineral and other deposits are zoned so that the metal grades vary, and therefore the economics of exploitation also varies. A threshold must be established around the ore that is economic to mine. This is the cut-off grade. The lowest cut-off grade is that at which the operating costs are covered.

© Springer International Publishing AG, part of Springer Nature 2019
F.-W. Wellmer et al., *Raw Materials for Future Energy Supply*,
https://doi.org/10.1007/978-3-319-91229-5

Fracking Fracking (hydraulic fracturing) is a hydraulic stimulation process that uses high fluid pressures to fracture low-permeability rocks and thereby, for example, recover the hydrocarbons contained in the rocks. The technology is used in drill holes for hydrocarbonproduction and deep geothermal energy. There are significant differences between the various applications, for example in the composition and the volumes of the fracking fluids that are used.

Good governance Good governance addresses the importance of the government responsibility, or the leadership with respect to the social common good. A comprehensive understanding of good governance extends beyond the government actions, and includes the relationship to state and non-state organizations. The term "governance" includes the way decisions are made by the state, and how political intentions are formulated and implemented. In other words, applying all the control measures to identify and implement collective technical and social obligations and, at the same time, referring to politics, business, science and society. While the term "government" focusses on the governing agents, the term "governance" refers to the processes for reaching its objectives and the relationships between the different organizations that are involved— government officials, industrial managers, scientists and the civil society (Compare Renn 2015: in part inspired by the analyses of the ESYS project working group). Good governance requires the creation of a political milieu in which a social ecological and economic development can succeed.

Herfindahl-Hirschmann Index (HHI) The Herfindahl-Hirschmann Index is an indicator for quantifying the concentration. The sales of a product are distributed among the number of relevant producers, normally not evenly. The resulting concentration of the global production can be calculated by squaring the market share of each producer and then calculating the sum of all these values. In the economic sciences, it is usual to present the results not as percent, but in values between 0 and 10,000 whereby 10,000 denotes a monopoly and values greater than 2500 are critical.

Lead time The lead time of a mining project is the time required from the discovery of the deposit to the commencement of production. Currently this is on average about ten years. Economic (price decrease or increase), technical (additional environmental requirements, improved exploitationtechnologies), or social issues (resistance from the local population) can impact the lead time. In some circumstances, the development of a mining project can also be completely interrupted.

Peak The term "resourcepeak" refers to the maximum feasible production for a raw material based on the available occurrences of the raw material. The term is essentially based on the finite number of crude oil occurrences that was predicted by K.M. Hubbert in 1956, or the so-called "Peak Oil", and has subsequently been used for other natural resources, especially the mineral resources, such as phosphorus.

Primary production of natural resources This term pertains to the exploitation of natural resources by mining. It is distinguished from secondary production of raw materials.

Reserves, resources, geopotential Reserves are known mineral occurrences that can be mined economically in the prevailing circumstances. Reserves can be increased by new discoveries or decreased as a result of, for example, a decline in the commodity prices. Resources are known mineral occurrences that cannot be mined economically in the prevailing circumstances. Geopotential includes occurrences that have not yet, or have only partly been discovered, and it is therefore not possible to make any statement on the possibilities for economic exploitation. The geopotential is therefore the big unknown factor. All three classes are dynamic, and are permanently changing. Their movements depend on, for example, the prevailing economic situation (including prices), technical advances or environmental requirements.

Secondary production of resources The secondary production of resources is the recovery of raw materials from discarded materials, such as products (vehicles, computers) and infrastructure (roads, transmission lines) that are deposited as waste and scrap in the technosphere.

"Shelved" mineral deposits The term "shelved" is used if events (usually a deterioration of the economic outlook or negative results from the deposit itself) occur during the exploration or preparation of a feasibility study of a mineral deposit that cause these works to be interrupted. The deposit is then set aside, or "shelved", but can always be available for revival. The lead time to the commencement of production for a shelved project can, under certain circumstances, be much shorter than that for a new mining project.

Sustainable resource supply The sustainable supply of natural resources means that the resource is exploited with the minimum possible impacts on the environment and social structure. During mining, or the primary exploitation of raw materials, which is essential for our supply with natural resources, it is not possible to avoid some impacts particularly on the environment. However, it is possible to take measures such as environmental requirements to minimize these impacts. According to sustainability principles, the people that are affected by the mining should be included in the decision-making process and should have a share in the profits. The economic aspect of sustainability, whereby the mining should not only benefit individual organizations or companies but should also serve the common good of the people affected by the resourceproduction, is therefore also addressed.

Technosphere The technosphere, or anthroposphere, are synonyms for the zone of human activity. They incorporate everything that has been created by humans, and include for example the mass products such as vehicles or electronic items, infrastructure such as constructions and supply systems, machines, but also waste and miningdumps.

The feedback control cycle of raw material ty supply The feedback control cycle of raw materialsupply describes the market mechanisms on the supply- and demand-side that are initiated by price indictors or scarcity of resources to regain a new market equilibrium.

Weighted country risk The weighted country risk for a primary raw material or an intermediate product is the estimated risks for each of the source countries, which are determined from the World Governance Index, weighted with the proportion each country contributes to the global production.

World Governance Index (WGI) The World Governance Index of the World Bank is an index derived for every country from six evaluation indicators: (1) "voice and accountability" measures the degree to which the people of a country are able to participate in the election of their government, and also includes factors related to freedom of speech, press and assembly; (2) "political stability and absence of violence" expresses the probability for destabilizing the government by means of unconstitutional or violent means (including terrorism); (3) "government effectiveness" evaluates the quality of the public services and administration as well as their independence from political interference; (4) "regulatory quality" evaluates the ability of the government to enable the development of the private sector by enacting laws and regulations; (5) "rule of law" evaluates the confidence in and the compliance of regulations in the society. This also covers the enforcement of contracts and property rights. The quality of the courts and police as well as the probably of crime and violence are also included; (6) "control of corruption" comprises the extent to which the public institutions are controlled by the interests of private profit and includes all scales of corruption as well as the appropriation of the state by elites and private interests. In general, the lower the level in the World Government Index, the higher the country risk.

Index

Italicsed cases refers to singular/plural terms and bold cases refers to proper nouns, abbreviations and acronyms

A
Abbreviation, 7
Ablated, 87
Absorption, 124
Abstract, 60, 95
Acatech, 120
Acceptance, 3, 60, 64–66, 70, 176
Accumulator, 148, 149
Acquisition, 50–52
Active cycle, 82
Active economic cycle, 80, 82
Add-on project, 76
Adelphi, 152
Adenosine Triphosphate (ATP), 136
Adjustment, 152, 159, 160
Adjustment clause, 173
Administration, 69
Advanced technology products, xvii
Advantage, 18, 52, 64, 77, 108, 122, 123, 138, 164, 171, 173, 175
Advisory committee, 16
Aerospace, 71
Affinities, 89
Affluence, 60, 127
Agrarian area, 127
Agrarian bioenergy, 175
Agrarian biomass, 132, 175
Agrarian biomass waste, 175
Agrarian waste, 132, 138
Agreement, 170
Agricultural biomass, 124, 132
Agricultural country, 110
Agricultural productivity, 135, 175
Agriculture, 40, 61, 124, 127–129, 131, 132, 134, 135, 139, 154, 175

Airborne, 71, 73
Airplane, 138
Airship, 162
Aitik Mine, 74
Alberta, 69
Alberta Heritage Savings Trust Fund, 69
Alert, 19
Algeria, 121, 163
Alliance for Securing Natural Resource, 113
Alternative, 1, 19, 55, 65, 93, 158, 159, 162, 165, 171, 173, 175
Alternative strategies, 55, 59, 170
Aluminium, 82
Alumino-borosilicate, 45
Aluminum, 6, 27, 38, 40–43, 79, 81, 85, 91, 111, 150, 164
Ammonium, 135
Amorphous silicon, 184
Amortization, 32
Analytical technique, 73
Andina, 74
Animal, 128, 132, 133, 138
Animal feed, 107, 125, 132, 136
Animal product, 132, 133
Animal species, 132
Anthropogenic, 1, 116, 133, 136
Anthropogenic cycle, 165
Anthroposhere, 199
Anticyclical, 113
Antimony, 7, 12, 27, 34, 53, 59, 60, 91, 152, 158
Anxiety, 62
Appliances, 79, 80, 83, 84, 87, 89, 90, 95
Application for electricity, 153
Applied research, 96

© Springer International Publishing AG, part of Springer Nature 2019
F.-W. Wellmer et al., *Raw Materials for Future Energy Supply*,
https://doi.org/10.1007/978-3-319-91229-5

Printed in the United States
By Bookmasters